Best of the Brain from Scientific American

BEST OF THE Brain

from Scientific American

EDITED BY FLOYD E. BLOOM, M.D.

DANA
PRESS

NEW YORK
WASHINGTON, D.C.

Copyright 2007, all rights reserved
Published by Dana Press
New York/Washington, D.C.

DANA
PRESS

The Dana Foundation
745 Fifth Avenue, Suite 900
New York, NY 10151

900 15th Street NW
Washington, DC 20005

DANA is a federally registered trademark.

ISBN-978-1-932594-22-5

Library of Congress Cataloging-in-Publication Data
Best of the brain from Scientific American / edited Floyd E. Bloom.
    p. cm.
  ISBN-13: 978-1-932594-22-5
   1. Cognitive neuroscience. 2. Brain. 3. Neurophysiology. I. Bloom, Floyd E.
II. Scientific American.
  QP360.5.B52 2007
  612.8'233—dc22

                                                    2007015128

Cover Design by Christopher Tobias
Text Design by Richard Hendel
Compositor: Newgen
"The New Science of Mind" is reprinted by permission of Eric R. Kandel.
www.dana.org

TABLE OF CONTENTS

Introduction

FLOYD E. BLOOM, M.D.

The study of the human brain and its disease remains one of the greatest scientific and philosophical challenges ever undertaken. The knowledge we have collected has become so plentiful and significant that it tends to intimidate novices and tempt the unaware into believing that the most important findings are behind us. In fact, the past accomplishments now allow us to pose better and still more interesting questions.

The discoveries that defined the 1990s—designated the "Decade of the Brain" by the Library of Congress and the National Institute of Mental Health, and proclaimed as such by President George H. W. Bush—have given rise to findings in this young millennium that span disciplines and have far-reaching implications. The articles in this collection reflect the promise, excitement and intrigue in many areas since the official end of the Decade of the Brain.

We begin with a section that reveals how researchers on the frontiers of brain science have come to focus on the biological mechanisms underlying mental activity in normal, experimental or clinical conditions. The articles under the heading "Mind" deal with behavior, personality and cognition. They focus on what are sometimes referred to as the "higher brain functions," for which the underlying molecular and cellular operations remain a bit murky. But the mental operations that these stories explore are tantalizingly familiar to us all.

We have chosen "Matter" for the title of the largest section in the book because diseases of the brain often reveal the holes in our knowledge of the healthy brain, and those illnesses that remain untreatable hold the greatest challenge for the researchers who study them. The articles in this section explain how scientists are reviewing the origins of the thought and emotional problems that characterize schizophrenia and depression, and how the mental activities of the brain emerge from the brain's underlying biological organization.

"Vision: A Window on Consciousness" and "Rethinking the 'Lesser' Brain" tackle our very concepts of higher-order mental functions. What we mean by the term "consciousness" can often be most simply approached by analysis of the world—objects and individuals we see around us, who or what they are, how we have interacted with them in the past—and how that past practice may need to be altered in the present context. The "lesser" brain, or cerebellum, has classically been thought to have responsibility as a motor coordination region, monitoring the locations of the limbs and trunk relative to the body and to the pull of gravity, so that a decision to adjust, move, reach or walk can be executed smoothly and efficiently. Recent research, however, suggests that the cerebellum has much more responsibility than that and, in fact, participates in several aspects of learning.

Although the pace of progress can seem excruciatingly slow for people with disorders of thought and emotion and their families, this too is an area where work is progressing. The articles "Diagnosing Disorders," "Turning Off Depression," "The Addicted Brain," and "Decoding Schizophrenia" highlight what draws the line between modulations of normal behaviors and those that physicians diagnose as diseases of the brain. In some cases, when the abnormal behaviors come and go, as they sometimes do in schizophrenia and often do in depression, judging the effectiveness of medications can be problematic in any individual patient. In order to develop therapies that are more effective than those currently available, scientists continue to probe for the underlying biological differences in the brains of these patients, research that provides the only avenue for new medications to emerge. In individuals who become dependent on drugs of abuse, legal and illicit, the brain adapts to the initially rewarding effects of the drug, and the wearing-off of those effects then motivates continued seeking and consumption of drugs. Treating this vicious cycle is now becoming a very realistic opportunity.

Theoretical studies that seek to identify the principal features of complex sensory discriminations or of information transfer lay the foundation for the "Tomorrow's Brain" section, which includes articles that address important advances in areas ripe for further study. "Neuromorphic Microchips" describes efforts to understand the structure and function of the intricate sensory detectors of the eye and ear that are leading to the production of nanodevices to restore vision and audition. "The Coming Merging of Mind and Machine" reveals similar efforts to read the combined activity of large ensembles of neurons engaged in planning limb movements, efforts that are leading to the development of a brain interface with machines. "Controlling Robots with the Mind" offers a look at the potential of this technology to restore movement and the ability to communicate despite the ravages of disease upon brain tissue. Yet to come from the brain sciences are profound insights from the substantial and growing research on the "systems" of the brain, the interconnected ensembles of neurons that govern the sensory, motor, and other integrative functions that regulate breathing, eating, drinking and hormonal cycles.

Together, the articles in this final section shine a bright light on the threshold that neuroscience is now crossing. Major separate communities dedicated to the study of brain development in a variety of organisms, as well as to the process by which the aging brain becomes senescent, are now viewing each other with new appreciation and starting to come together in pursuit of common quarries. For example, these disparate fields have been linked recently by the study of substances that can initiate programmed neuronal death, or apoptosis, and study of the equally robust phenomenon of activity-dependent modifications to the brain's structural and functional abilities, including the striking but limited ability to form new neurons long after brain development was thought to have been completed.

Although these articles are not only approachable but fascinating to the general reader, an introduction to the specialized terminology of the brain is important for deeper understanding. Beginning students of brain sciences could justifiably find themselves confused if they are not familiar with this terminology, for descriptions of the nervous system of virtually all organisms ever investigated use multiple overlapping and often synonymous terms.

In vertebrates, the pioneers of brain research named the components of the nervous system for both their appearance and their location. For example, the names of the major parts of the brain resulted from creative interpretations by early dissectors of the brain, attributing names to brain segments based on their appearances in the freshly dissected state: hippocampus (shaped like the sea horse) or amygdala (shaped like the almond), cerebrum (the main brain) and cerebellum (the small brain).

Although the brain is a single, unique organ, it is more properly conceptualized as an assembly of interrelated neural systems that regulate their own and each other's activity in a dynamic, complex fashion largely through a chemical process among cells called neurotransmission. The brain and spinal cord together constitute the central nervous system, or CNS. The two cerebral hemispheres make up the largest division of the brain. Regions of the cortex are classified in several ways. One category comprises the types of information processed, such as *sensory* (somatosensory, visual, auditory and olfactory), *motor* (walking, reaching, writing), or *associational* (considered to underlie higher cortical functions such as abstract thought, memory and consciousness). A second category turns to anatomical position in the brain (frontal, temporal, parietal and occipital), and a third focuses on the geometrical relationships between the size, shape, and density of neurons in the major cortical layers.

Neurons communicate chemically through specialized contact zones called synapses in the central nervous system, or junctions in the peripheral nervous system, which comprises skeletal and smooth muscles and various secretory glands. For the most part, vertebrate nervous systems use a few specific molecular forms of neurotransmitters. The majority of synapses are mediated by simple amino acids, the chemical building blocks of proteins. In the nervous system certain amino acids such as glutamate (the major excitatory neurotransmitter) and gamma-aminobutyric acid (GABA, the major inhibitory neurotransmitter) act on their own as chemical signals across synapses. Other neurotransmitters include acetylcholine and the monoamine transmitters, serotonin, dopamine, norepinephrine, and epinephrine.

The importance of neurotransmitters lies in the fact that virtually every drug used to treat diseases of the brain augments, blunts, or in some other discrete manner modifies the effects of one or more neurotransmitters. Whether drugs of the future will expand these therapeutic mechanisms remains to be seen, as addressed in "Treating Depression: Pills or Talk" and "The Quest for a Smart Pill."

In many cases, synapses in both the central nervous system and at the peripheral nerve junction have been found to employ more than one transmitter, in which case the signals provided by the amino acid or amine are supplemented by chemically more complex transmitters called neuropeptides. The specific proteins of synapses are essential for neurotransmitter transmitter synthesis, storage and metabolism. Proteins also play a role in the activity-dependent release of the transmitter by the "sending" neuron and the response to the transmitter by the "receiving" neuron.

So far, we have emphasized the operations of the brain at only one conceptual level, namely the cellular operations in which one neuron speaks to another neuron. In a more formal view, four research strategies actually provide the neuroscientific bases of neuropsychological phenomena: molecular, multicellular (or systems) and behavioral, in addition to the aforementioned cellular approach. The intensively exploited molecular level has been the traditional focus for characterizing drugs that alter behavior. Thus, the most basic cellular phenomena of neurons now can be understood in terms of discrete molecular entities, including precisely how the four major mineral ions of the brain regulate the excitability of individual neurons, and their ability to send selective signals quickly over long distances.

An understanding at the systems level is required to assemble the descriptive structural and functional properties of specific central transmitter systems, linking the neurons that make and release this transmitter to the possible effects of its release at the behavioral level. While many such transmitter-to-behavior linkages have been postulated, it has proved diffcult to confirm the essential involvement of specific transmitter-defined neurons in the mediation of specific mammalian behavior.

Research at the behavioral level often can illuminate the integrative phenomena that link populations of neurons into extended specialized circuits, ensembles, or more pervasively distributed functional sensory, motor and integrative systems noted previously that integrate the physiological expression of a learned, reflexive, or spontaneously generated behavioral response. The entire concept of "animal models" of human psychiatric diseases rests on the assumption that scientists can appropriately infer from observations of animals' behavior and physiology (heart rate, respiration, locomotion, etc.) that the states they experience are equivalent to the emotional states human beings experience when we show similar physiological changes.

As intriguing, insightful, and heuristic as the products of the neuroscientific community have been until now, the most exciting vistas lie ahead. As the several whole-genome projects have now made it possible to use computer-based information to search, identify and compare multiple similarities, differences and even clusters of minor genetic differences, it will soon be possible to find those genetic differences that can modify the ways in which individuals react to the events of their external environment to learn and adapt—or in some cases fail to adapt and develop diseases. Genes that can modify these complex adaptations to environmental events may someday provide the most opportune methods to maintain health and avoid disease. Perhaps many of these discoveries will appear in time for the next collection of brain-related articles. Until then, I bid you good reading.

San Diego

September 2006

Part 1 Mind

Unleashing Creativity

Moments of brilliance arise from complex cognitive processes. Piece by piece, researchers are uncovering the secrets of creative thinking

ULRICH KRAFT

Jancy Chang, a high school art teacher in San Francisco, had been painting since she was a child. She varied her technique from Western-style watercolors to classical Chinese brushstrokes, but she always strove for realism: painting landscapes and people in social settings as literally as she could. Then, in 1986, at age 43, she began to have problems performing her job. Grading, preparing for class, putting together lesson plans—everything that she had previously done with ease—became increasingly difficult over the next few years. By 1995 she could no longer remember the names of her students and was forced to take early retirement.

Understandably frightened, Chang had started seeing neurologist Bruce L. Miller, clinical director of the Memory and Aging Center at the University of California at San Francisco. He diagnosed her with frontotemporal dementia. This relatively rare form of dementia selectively damages the temporal and frontal lobes, primarily in the brain's left hemisphere. These regions control speech and social behavior and are intimately involved in memory. Patients often become introverted, exhibit compulsive behaviors and lose inhibitions that would otherwise prevent them from acting inappropriately toward others in social settings.

Miller observed all these changes in Chang, but he also found that her creative powers were growing remarkably. "The more she lost her social and language abilities, the wilder and freer her art became," he notes. The same lack of inhibition that caused embarrassing moments in public allowed her to break the shackles of her realism art training and become increasingly impressionistic and abstract. Her paintings were much more emotionally charged.

Art teacher Jancy Chang sought realism in her own work, like *Jahua House*, but as dementia set in, her paintings became increasingly imaginative, like the wildly impressionistic *Four Masks*.

Miller was astonished. The last place he expected talent to bloom was in the brain of a person whose mental functions were deteriorating because of crumbling neurons. But it turned out that Chang was not an isolated case. Miller later identified other men and women whose latent creativity burst forth as frontotemporal dementia set in—even in patients who had little prior interest in artistic pursuits. One man, a stockbroker who had never before been touched by the muse, traded his conservative suits for the most radical styles he could find. He developed a passion for painting and went on to win several art prizes. Another person began to compose music even though he had no musical training. A third invented a sophisticated chemical detector at a stage when he could recall only one in 15 words on a memory test.

The ability to create is one of the outstanding traits of human beings. From harnessing fire to splitting the atom, an inexhaustible stream of innovative flashes has largely driven our social development. Significant insight into the neuronal mechanisms underlying the creative thought process is coming from work with patients who, like Chang, have suddenly acquired unusual skills as a result of brain damage. Using technical advances such as functional magnetic resonance imaging and electroencephalography, neuroscientists are trying to determine just where those sparks originate.

Scientific understanding of creativity is far from complete, but one lesson already seems plain: originality is not a gift doled out sparingly by the gods. We can call it up from within us through training and encouragement. Not every man, woman or child is a potential genius, but we can get the most out of our abilities by performing certain kinds of exercises and by optimizing our attitudes and environment—the same factors that help us maximize other cognitive powers. Some of the steps are deceptively simple, such as reminding ourselves to stay curious about the world around us and to have the courage to tear down mental preconceptions. Steven M. Smith, a professor of psychology at the Institute for Applied Creativity at Texas A&M University, says many people believe that only a handful of geniuses are capable of making creative contributions to humanity: "It just isn't true. Creative thinking is the norm in human beings and can be observed in almost all mental activities."

The ease with which we routinely string together appropriate words during a conversation should leave no doubt that our brains are fundamentally creative. What scientists are trying to discover is why the engine of inspiration seems to be always in high gear in some people while others struggle.

IT'S NOT INTELLECT

Intelligence is not a crucial ingredient. U.S. military leaders recognized this seeming contradiction more than 50 years ago. During World War II, the U.S. Air Force sought to identify fighter pilots who would be able to get out of jams in unorthodox ways. Officials wanted pilots who would not simply bail out in an emergency but who would be more likely to save themselves and their aircraft. Initially, military scouts used conventional intelligence tests to identify such candidates. But they soon realized a high IQ was useless in finding inventive superpilots, and they resorted to more anecdotal measures.

About the same time, psychologist Joy Paul Guilford of the University of Southern California noted that intelligence did not mirror the totality of a person's cognitive capacity. In the late 1940s Guilford developed a model of human intellect that formed the basis for modern research into creativity. A crucial variable is the difference between "convergent" and "divergent" thinking.

Convergent thinking aims for a single, correct solution to a problem. When presented with a situation, we use logic to find an orthodox solution and to determine if it is unambiguously right or wrong. IQ tests primarily involve convergent thinking. But creative people can free themselves from conventional thought patterns and follow new pathways to unusual or distantly associated answers. This ability is known as divergent thinking, which generates many possible solutions. In solving a problem, an individual proceeds from different starting points and changes direction as required, which Guilford explained leads to multiple solutions, all of which could be correct and appropriate.

Guilford tried to find a measurable "creativity quotient" analogous to IQ, but

STEPS TO A CREATIVE MIND-SET

Wonderment. Try to retain a spirit of discovery, a childlike curiosity about the world. Question understandings that others consider obvious.

Motivation. As soon as a spark of interest arises in something, follow it.

Intellectual courage. Strive to think outside accepted principles and habitual perspectives such as "We've always done it that way."

Relaxation. Take the time to daydream and ponder, because that is often when the best ideas arise. Look for ways to relax and consciously put them into practice.

his efforts and those of other researchers since his time have all failed. A few techniques, such as the Torrance Test of Creative Thinking, can give a sense of which people in a test group may be more creative. But deciding which of their many responses can be characterized as especially creative is simply too dependent on the personal judgment of the tester.

Rather than using a standardized test, today's creativity experts look for certain characteristics that people who excel at divergent thinking seem to exhibit. The following are prime examples:

Ideational fluency. The number of ideas, sentences and associations a person can think of when presented with a word.

Variety and flexibility. The diversity of different solutions a person can find when asked to explore the possible uses of, say, a newspaper or a paper clip.

Originality. The ability to develop potential solutions other people do not reach.

Elaboration. The skill to formulate an idea, expand on it, then work it into a concrete solution.

Problem sensitivity. The ability to recognize the central challenge within a task, as well as the difficulties associated with it.

Redefinition. The capacity to view a known problem in a completely different light.

LEFT OR RIGHT?

Guilford's distinction between convergent and divergent thinking prompted neuroscientists to examine whether the two processes took place in different brain regions. Their experiments, particularly those conducted in the 1960s by

TORRANCE TEST

In a standardized Torrance Test of Creative Thinking, subjects are given simple shapes (left column) and are asked to use them (top row) or combine them (middle row) in a picture or to complete a partial picture (bottom row). Evaluators judge whether the results are more or less creative.

	Starting Shapes	Completed Drawing	
		More Creative	Less Creative
Use	◯	Mickey Mouse	Chain
Combine	▯ ○ ▱ ▽	King	Face
Complete	⌴ °	A fish on vacation	Pot

psychobiologist Roger W. Sperry of the California Institute of Technology, revolutionized neurology and psychology. Sperry worked with so-called split-brain patients who suffered from epilepsy that did not respond to conventional medical treatment. The only way to end their horrible seizures was to surgically sever their corpus callosum, the fibrous structure that links the brain's left and right hemispheres.

Sperry and his colleague Michael Gazzaniga, now at Dartmouth College, put patients through a series of sophisticated experiments, which led to the breakthrough discovery that the left and right hemispheres do not process the same information. Sperry won the 1981 Nobel Prize in Physiology or Medicine for the work. Among other duties, the left hemisphere is responsible for most aspects of communication. It processes hearing, written material and body language. The right hemisphere processes images, melodies, modulation, complex patterns such as faces, as well as the body's spatial orientation.

The functional differences between the hemispheres are the subject of intense research today. Studies of stroke patients confirm the basic division of labor. Damage to the right hemisphere, for example, leaves speech largely intact but harms body awareness and spatial orientation. But researchers have noted another interesting correlation: patients with right hemisphere strokes lose whatever creative talents they had for painting, poetry, music, even for playing games such as chess.

The accumulation of experimental evidence now proves that the left hemisphere is responsible for convergent thinking and the right hemisphere for divergent thinking. The left side examines details and processes them logically and analytically but lacks a sense of overriding, abstract connections. The right side is more imaginative and intuitive and tends to work holistically, integrating pieces of an informational puzzle into a whole.

Consider a poem. When an individual reads it, his left hemisphere analyzes the sequence of letters and integrates them into words and sentences, following the logical laws governing written language. It checks for grammatical and morphological meaning and grasps the factual content. But the right hemisphere interprets a poem as more than a string of words. It integrates the information with its own prior ideas and imagination, allows images to well up, and recognizes overarching metaphorical meaning.

CREATIVITY UNLEASHED

The right hemisphere's divergent thinking underlies our ability to be creative. Curiosity, love of experimentation, playfulness, risk taking, mental flexibility, metaphorical thinking, aesthetics—all these qualities play a central role. But why does creativity remain so elusive? Everyone has a right hemisphere, so we all should be fountains of unorthodox ideas.

Consider that most children abound in innovative energy: a table and an old blanket transform into a medieval fortress, while the vacuum cleaner becomes the knight's horse and a yardstick a sword. Research suggests that we start our young lives as creativity engines but that our talent is gradually repressed. Schools place overwhelming emphasis on teaching children to solve problems correctly, not creatively. This skewed system dominates our first 20 years of life; tests, grades, college admission, degrees and job placements demand and reward targeted logical thinking, factual competence, and language and math skills— all purviews of the left brain. The propensity for convergent thinking becomes increasingly internalized, at the cost of creative potential. To a degree, the brain is a creature of habit; using well-established neural pathways is more economical than elaborating new or unusual ones. Additionally, failure to train creative

INHIBITION LOST

When brain tissue in the frontotemporal lobes atrophies, typically because of dementia, victims often lose their inhibitions. This change can lead to increasingly inappropriate social behavior, such as loud outbursts or sexual references. The lack of self-control can also markedly enhance creative thinking and talents such as painting and sculpture. Vincent van Gogh fit this profile perfectly late in his career; this is a work of his from 1888, two years before his death.

faculties allows those neural connections to wither. Over time it becomes harder for us to overcome thought barriers. Creativity trainers like to tell clients: "If you always think the way you always thought, you'll always get what you always got—the same old ideas."

Bruce Miller's examination of Jancy Chang and other patients like her lends credence to the notion that the logical left hemisphere may block the creative right side. With the help of imaging techniques, Miller has determined that people with frontotemporal dementia lose neurons primarily in the left hemisphere. Patients have trouble speaking and show no regard for social norms.

And yet this very lack of inhibition allows dormant artistic talents to bloom. Miller draws parallels to creative geniuses such as Vincent van Gogh and Francisco Goya, who ignored social expectations and developed unorthodox styles that opposed contemporary conventions. Great artists often exhibit an ability to transcend social and cognitive walls.

Nevertheless, it is wrong to assume that the left hemisphere is all that stands in the way of genius. Not every unconventional idea is necessarily a good one; many completely miss a problem at hand or are simply outlandish. The most important creative work is useful, relevant or effective. And it is the left hemisphere that conducts this self-evaluation as creative thoughts bubble up from the right. As Ned Herrmann, artist, actor, management trainer and author of *The Creative Brain* (Ned Herrmann Group 1995), notes, the left brain keeps the right brain in check. Creativity involves the entire brain.

VOYAGE OF DISCOVERY

Convergent thinking is also required for a creative breakthrough. Inspirational thunderbolts do not appear out of the blue. They are grounded in solid knowledge. Creative people are generally very knowledgeable about a given discipline. Coming up with a grand idea without ever having been closely involved with an area of study is not impossible, but it is very improbable. Albert Einstein worked for years on rigorous physics problems, mathematics and even philosophy before he hit on the central equation of relativity theory: $E = mc^2$. As legendary innovator Thomas A. Edison, author of 1,093 patents, noted drily, "Genius is 1 percent inspiration and 99 percent perspiration."

Various psychologists have floated different models of the creative process, but most involve an early "preparation" phase, which is what Edison was talking about. Preparation is difficult and time-consuming. Once a challenge is identified, a person who wants to solve it has to examine it from all sides, including new perspectives. The process should resemble something like an intellectual voyage of discovery that can go in any direction. Fresh solutions result from disassembling and reassembling the building blocks in an infinite number of ways. That means the problem solver must thoroughly understand the blocks.

Smith of Texas A&M emphasizes how important it is to be able to combine ideas. He says people who are especially inventive have a gift for connecting elements that at first glance may seem to have nothing in common. To do that, one must have a good grasp of the concepts. The more one knows, the easier it will be to develop innovative solutions.

In this context, psychologist Shelley H. Carson of Harvard University reached an interesting insight in 2003. She analyzed studies of students and found that those who were "eminent creative achievers"—for example, one had published a novel, another a musical composition—demonstrated lower "latent inhibition" on standard psychological tests than average classmates. Latent inhibition is a sort of filter that allows the brain to screen out information that has been shown by experience to be less important from the welter of data that streams into our heads each second through our sensory system. The information is cast aside even before it reaches consciousness. Think about your act of reading this article right now; you have most likely become unaware that you are sitting in a chair or that there are objects across the room in your peripheral vision.

Screened data take up no brain capacity, lessening the burden on your neurons. But they are also unavailable to your thought process. Yet because creativity depends primarily on the ability to integrate pieces of disparate data in novel ways, a lower level of latent inhibition is helpful. It is good to filter out some information, but not too much. Then again, lower latent inhibition scores have been associated with psychosis.

Latent inhibition has a corollary: too much specialized knowledge can stand in the way of creative thinking. Experts in a field will often internalize "accepted" thought processes, so that they become automatic. Intellectual flexibility is lost. For example, a mathematician will very likely tackle a difficult problem in an analytical way common to her professional training. But if the problem resists solution by this method, she may well find herself at a mental dead end. She has to let go of the unsuitable approach.

THE BATHTUB PRINCIPLE

Letting go to gain inspiration may be difficult. One aid is to simply get away from the problem for a while. Creativity does not prosper under pressure. That is why so many strokes of genius have occurred outside the laboratory, in situations that have nothing to do with work. Legend has it that Greek mathematician and mechanical wizard Archimedes was stepping into a bathtub when the principle of fluid displacement came to him—the original "eureka!" moment. Organic chemist Friedrich August Kekulé had a dream about snakes biting their own tails; his eureka moment occurred the next morning, when he depicted the chemical structure of benzene as ring-shaped.

Creative revelations come to most people when their minds are involved in an unrelated activity. That is because the brain continues to work on a problem once it has been supplied with the necessary raw materials. Some psychologists call this mental fermentation or incubation. They surmise that associative connections between ideas and imagination that already exist in the mind become weaker and are transformed by new information. A little relaxation and distance change the mind's perspective on the problem—without us being aware of it. This change of perspective allows for alternative insights and creates the pre-conditions for a fresh, and perhaps more creative, approach. The respite seems to allow the brain to clear away thought barriers by itself. At some point, newly combined associations break into consciousness, and we experience sudden, intuitive enlightenment.

The little insights and breakthroughs we all experience should encourage us to believe that bigger eureka moments are possible for anyone. Our brains bestow moments of illumination almost as a matter of course, as long as there has been adequate preparation and incubation. The catch is that because the neural processes that take place during creativity remain hidden from consciousness, we cannot actively influence or accelerate them. It therefore behooves even the most creative among us to practice one discipline above all—patience.

MORE TO EXPLORE

Carson, S. H., J. B. Peterson, and D. M. Higgins. 2003. Decreased latent inhibition is associated with increased creative achievement in high-functioning individuals. *Journal of Personality and Social Psychology* 85 (3): 499–506.

Ward, T. B., R. A. Finke, and S. M. Smith, 2000. *Creativity and the mind: Discovering the genius within.* Perseus Publishing.

Zeki, Semir. 2001. Artistic creativity and the brain. *Science* 293 (July 6): 51–52.

ULRICH KRAFT is a physician and medical writer based in Fulda, Germany. Originally published in *Scientific American Mind*, Vol. 16, No. 1, April 2005.

2

Stimulating the Brain

Activating the brain's circuitry with pulsed magnetic fields may help ease depression, enhance cognition, even fight fatigue

MARK S. GEORGE

Bleary-eyed, the pilot stares at the instruments while sipping stale coffee. The cup is nearly empty, as is the radar screen. So, he realizes, are the airplane's fuel tanks, not to mention his own energy reserves. Another cup certainly won't help much. His co-pilot dozes beside him, having already flown several legs of their long mission to deliver sorely needed humanitarian aid to the other side of the world. The pilot considers, then rejects, popping a pep pill. Uppers make him jumpy, a bad feeling to have during the tricky nighttime aerial-refueling maneuver he will soon have to execute. Suddenly the radar shows a blip orbiting up ahead. Scanning the cloudy sky for the tanker's navigation lights, the pilot knows he has to get focused fast. He flips a switch. A "rat-a-tat-tat" sound, like that of a staple gun, echoes through his helmet, and fatigue abruptly flees his mind. Clear-headed for the first time in what seems days, the pilot almost immediately spies lights flashing in the murky distance. He nudges the co-pilot, who absently toggles his own switch as he stifles a yawn. Muffled snapping noises follow. Fully awake, the aviators steer for the flying gas station circling overhead.

In the scenario above, sharp sounds emerge when electromagnets inside the helmets generate magnetic fields to excite particular parts of the pilots' brains—areas that govern tiredness and wakefulness. Neuroscientists developing this novel noninvasive technique call it transcranial magnetic stimulation (TMS). TMS employs head-mounted wire coils that send strong but very short magnetic pulses directly into specific brain regions, thus safely and painlessly inducing tiny electric currents in a person's neural circuitry.

OVERVIEW/ELECTROMAGNETIC EXCITATION

- Neuroscientists utilize a variety of electromagnetic stimuli to directly activate neurons in the brain. Electroconvulsive therapy (ECT), a procedure in which electrodes are attached to the scalp, is the best-known of these techniques. Its use, however, remains somewhat problematic for various reasons.
- For a decade, researchers have been experimenting with pulsed magnetic fields that induce electrical activity in specific areas of the brain safely and painlessly. The ability of transcranial magnetic stimulation (TMS) to target specific brain regions is key to many new applications.
- TMS offers potential treatments for depression and other neuro-physiological disorders. The technology may also provide a nonpharmaceutical method to rouse people from the effects of severe fatigue or to teach them a new skill.

This scenario is still speculative, but research to make this promising technology a reality is advancing steadily. The Defense Advanced Research Projects Agency (DARPA) is funding several studies to investigate the use of TMS to improve the performance of U.S. service personnel exhausted by protracted field operations. And DARPA is not alone in its interest in TMS, because the procedure offers one of the most promising technological (nonpharmaceutical) methods to literally turn on and off particular parts of the human brain.

Some TMS researchers, for example, are inducing temporary brain "lesions" in healthy subjects to gain insight into fundamental neuronal mechanisms such as speech and spatial perception: they inhibit a basic brain function with a magnetic pulse stream and then compare the "before" condition with the "after." Other investigators are trying to determine whether the hyperactive brain regions that create epileptic seizures might be quieted with magnetic fields. Still other neuroscientists are attempting to employ TMS to alter the operation of specialized nerve cell networks to enhance people's memory and learning. Many of my colleagues are looking for ways to use the technology as an alternative to seizure-causing electroconvulsive therapy (ECT) to ease depression. Whatever the goals, TMS holds great potential as a tool for understanding how the brain works, correcting its dysfunctions and even augmenting its abilities.

THE ELECTRIC BRAIN

TMS takes advantage of the fact that the brain is fundamentally an electrical organ that transmits electrical signals from one nerve cell to the next. When a TMS coil is activated near the scalp, an extremely powerful and rapidly changing magnetic field travels unimpeded through skin and bone. Although the field reaches a strength of nearly 1.5 tesla—tens of thousands of times that of the earth's magnetic field—each pulse lasts for less than a millisecond. The popping sound it generates when it is operating arises from the passing of current through the insulated coil.

In the brain, the magnetic field encounters resting nerve cells and induces small electric currents to flow in them. Thus, electrical energy in the copper-wire coil (typically encased in a paddlelike wand) is converted into magnetic energy, which is then changed back into electric current in the neurons of the brain. The $30,000 to $40,000 TMS machines are manufactured by the Magstim Company Limited in Whitland, Wales, by Dantec/Medtronic in Denmark and in Shoreview, Minn., and by Neuronetics in Malvern, Pa.

Unlike purely electrical techniques—such as ECT and others, which involve attaching electrodes to the scalp or even to brain or nerve tissue—TMS creates a magnetic field that enters the brain without any interference or direct contact. The technique can be thought of as electrodeless electrical stimulation. Although magnetism does interact with biological tissue to some degree, the majority of TMS effects most likely derive not from the magnetic fields directly but from the electric currents they produce in neurons.

MAGNETIC EXCITATION

The idea of using electromagnetic fields to alter neural function goes back to at least the early 1900s. Psychiatrists Adrian Pollacsek and Berthold Beer, who worked down the street from Sigmund Freud in Vienna, filed a patent to treat depression and neuroses with an electromagnetic device that looked surprisingly like a modern TMS apparatus.

Today's TMS technology took shape in 1985, when medical physicist Anthony T. Barker and his colleagues at the University of Sheffield in England created a focused electromagnetic device with enough power to create currents in the spinal cord. They quickly realized that their equipment could also directly and noninvasively stimulate the brain itself. Since then, the field of TMS research has exploded.

Unfortunately, TMS devices can excite only the surface cortex of the brain because magnetic field strength falls off sharply with distance from the coil

TRANSCRANIAL MAGNETIC STIMULATION

Localized brain-cell excitation results from the use of a transcranial magnetic stimulation (TMS) machine. When researchers operate a TMS coil near a subject's scalp, a powerful and rapidly changing magnetic field passes safely and painlessly through skin and bone. Each brief pulse, lasting only micro-seconds, contains little energy. Because the strength of the magnetic field falls off rapidly with distance, it can penetrate only a few centimeters to the outer cortex of the brain (top right). On arrival, the precisely located field induces electric current in nearby neurons, thus activating targeted regions of the brain (bottom right). A principal benefit of TMS is that it requires no direct electrical connection to the body, as is required for electroconvulsive therapy.

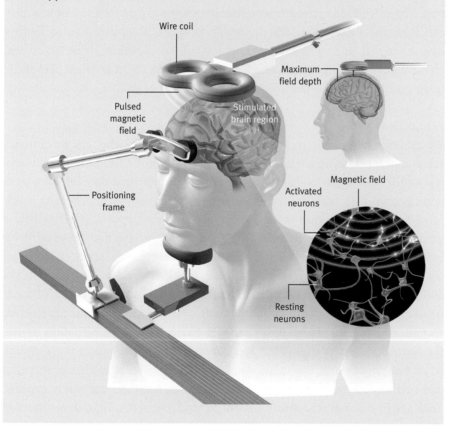

Wire coil

Maximum field depth

Pulsed magnetic field

Stimulated brain region

Positioning frame

Magnetic field

Activated neurons

Resting neurons

(maximum range: two to three centimeters). A magnetic field that can safely penetrate and activate the brain's central structures continues to be the Holy Grail of TMS research because it offers the possibility of treating difficult conditions such as Parkinson's disease.

When researchers send a single magnetic pulse into the motor cortex of a subject's brain, it produces a jerk in the hand, arm, face or leg, depending on where the coil is placed. One pulse directed to the back of the brain can generate a flash of light in the eyes. That is the extent of the immediate effects of single-pulse TMS, however. Magnetic field pulses emitted in rhythmic succession, which neuroscientists call repetitive TMS, or rTMS, though, can induce behaviors not seen with the use of single pulses. These results are now the subject of intense study.

For brief periods during stimulation, rTMS can block or inhibit a brain function. Repetitive TMS application over the speech-control motor area, for instance, can leave the subject temporarily unable to speak. Cognitive neuroscientists have employed this so-called functional knockout capability to reexplore and confirm our knowledge about which part of the brain controls which part of the body, insights that have been gleaned from decades of studying stroke patients.

FIELD LEARNING

When single nerve cells are made to discharge repeatedly, they can form themselves into functioning circuits. Researchers have found that stimulating a neuron with a low-frequency electrical signal can produce what they call long-term depression (LTD), which diminishes the efficiency of the intercellular links. High-frequency excitation over time can generate the opposite effect, which is known as long-term potentiation (LTP). Scientists believe that these cell-level behaviors are involved in learning, memory and dynamic brain changes associated with neural networks. The chance that one could use magnetic brain stimulation to alter brain circuitry in a manner analogous to LTD or LTP fascinates many researchers. Although this controversial notion remains unresolved, several studies have shown nerve cell network inhibition or excitation lasting for up to a few hours after rTMS application. The implications of these results could be enormous. If one could employ rTMS techniques to change learning and memory by resculpting brain circuits, the possibilities are nearly endless. TMS might be used on a stroke patient to teach the remaining, intact parts of the brain to pick up the functions formerly conducted by the damaged region. Or overactive brain circuits that drive epilepsy might be toned down, resulting in fewer seizures.

ZAPPING THE BRAIN

Whereas transcranial magnetic stimulation of the human brain came into modern use in the 1980s, electrical excitation has been around for at least a century. David Ferrier and others in the 1880s showed that direct electrical brain stimulation could change behavior and that activation of specific regions correlated with certain behavioral changes. For the past 100 years, neurosurgeons have stimulated the brain electrically during brain surgery, cataloguing the resulting effects along the way.

Physicians have long known that electrical stimulation could be therapeutic as well. During electroconvulsive therapy (ECT), a doctor applies electrodes directly to the scalp of an anesthetized subject with the goal of inducing a generalized seizure. For reasons that are still unclear, repeated ECT sessions over the course of several weeks are an effective treatment for depression, mania and catatonia. The technique is, however, associated

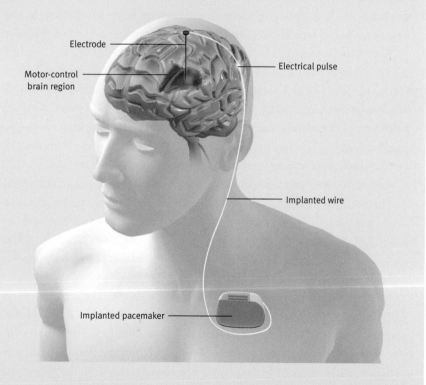

Persistent shaking (dyskinesia) caused by Parkinson's disease can be lessened by electrical deep brain stimulation, imparted by implanted electrodes.

with memory loss and requires repeated general anesthesia. Because the skull acts as a large resistor that spreads direct electric current, ECT cannot be focused on or directed to specific targets within the brain.

Of late, neuroscientists have explored other methods to electrically stimulate the brain. These new techniques tend to be either more focused or less invasive, or both, than the older ones. Employed in conjunction with the advanced brain-imaging technologies developed over the past two decades, these approaches are being used to build on our recently assembled understanding of how the brain works.

Two direct electrical brain-stimulation techniques have been approved for therapeutic use. In deep brain stimulation (DBS), a neurosurgeon guides a small electrode into the brain through a small hole in the skull with the help of three-dimensional images. The surgeon then connects the electrode to a pacemaker (signal generator) implanted in the chest. The pacemaker sends high-frequency electrical pulses directly into the brain tissue. DBS is approved by the U.S. Food and Drug Administration for the treatment of Parkinson's disease, typically in patients who no longer respond to medication.

Within the motor-control circuitry of the brain, several regions (including the internal globus pallidus, thalamus and subthalamic nucleus) are inhibitory in function and so act as brakes on movement. In current practice, neurosurgeons place DBS electrodes in those regions and then stimulate them at high frequencies to arrest the shaking (dyskinesia) that characterizes Parkinson's. The technique is being explored as a treatment for depression as well. Little information exists concerning what happens when DBS is applied to other brain regions or when low-frequency pulses are used.

Theoretically, DBS electrodes can be removed with no lasting damage. Thus, the procedure represents an advance over traditional ablative brain surgery in which neural tissue is lost forever. In rare cases, DBS can, however, lead to infections, strokes and even death, so it is largely restricted to patients who have failed to respond to other therapies.

Another electrical brain-excitation technique now in use is vagus nerve stimulation (VNS). The vagus nerve is an important cranial nerve that connects the brain with the body's viscera. Eighty years of research have shown that stimulation of the vagus nerve in the chest or neck can alter the operation of brain regions involved in the control of bodily functions. In the 1980s Jake Zabara of Temple University discovered that excitation of the vagus nerve could abort a seizure occurring in a dog. This finding led to clinical trials of the technique and eventual FDA approval of VNS for suppressing seizures.

Surgeons typically wrap the VNS electrode around the left vagus nerve in the neck and connect it to a pacemaker they have implanted in the patient's chest wall. The VNS apparatus can be programmed to produce electrical stimuli in various intermittent patterns.

Researchers are now conducting studies to determine if VNS has therapeutic value for other disorders such as depression and anxiety. As with the other stimulation approaches, we do not know if changing how the VNS electrical signals are delivered would produce different brain effects. Our group at the Medical University of South Carolina has investigated VNS within a functional magnetic resonance imaging scanner to determine whether altering VNS parameters achieves different results. If it is confirmed, one might modulate brain regions by varying the VNS pulse pattern at the neck. No brain surgery would be needed.—M.S.G.

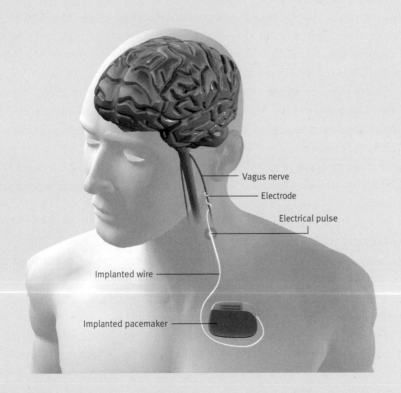

Vagus nerve

Electrode

Electrical pulse

Implanted wire

Implanted pacemaker

Direct electrical excitation of the vagus nerve in the neck can suppress the onset of brain seizures.

Recent experiments in our laboratory at the Medical University of South Carolina (MUSC) and elsewhere hint that rTMS might temporarily enhance cognitive performance, either during application or for short periods afterward. Investigators at the National Institute of Neurological Disorders and Stroke, for example, found that TMS applied to the prefrontal cortex can enable subjects to solve geometric puzzles more rapidly.

Most researchers working in this area stimulate subjects' brains over the prefrontal cortex or parietal cortex while they perform a task. To control for testing bias, neuroscientists also use deactivated ("sham") rTMS coils. Our lab is funded by DARPA to study whether rTMS might temporarily energize sleep-deprived individuals so they can perform better over the short term. Early results are promising. Another DARPA-supported group at Columbia University, led by Yaakov Stern and Sarah H. Lisanby, is exploring whether rTMS might be used to retrain subjects to accomplish a task in a different manner by shifting neural activity to an alternative cellular network that might be more resilient to stress or sleep deprivation.

Recent media reports have made public Australian claims that TMS might be used to unleash nascent savant skills (mastery of difficult tasks without training) in healthy subjects by temporarily disabling one brain hemisphere. This work has not yet been published in the scientific press. In fact, most neuroscientists believe that the reported effects are unlikely to be true. Researchers have supervised TMS sessions involving thousands of subjects and have yet to witness any so-called savant skill changes. Although existing artistic talents occasionally improve with the onset of dementia, we have not seen savant abilities emerge after TMS-like stimuli such as focal brain disability caused by trauma, stroke or surgery or after brain areas are injected with anesthetics.

WHAT EXCITES WHAT

Intriguing as these potential applications might be, they raise difficult questions. Scientists would like to ascertain exactly which neurons rTMS affects as well as the detailed neurobiological events that follow stimulation. In addition to figuring out which electromagnetic frequencies, intensities and dose regimens might produce different behaviors, researchers must decide (for each individual) exactly where to place the rTMS coil and whether to activate it when someone engages in a task. Scientists also need more knowledge about what rTMS is doing at both the cellular level—the effects on neurotransmitters, gene expression, synaptic changes—as well as at the circuit level.

DEEP BRAIN MAGNETIC STIMULATION

Transcranial magnetic stimulation fields extend only a few centimeters to the surface of the cortex. Although TMS is promising for certain applications, the procedure could find much wider use if it could reach to the central structures of the brain.

High-intensity TMS fields could penetrate farther into the brain, but they can cause seizures, tissue damage or discomfort. Thus, a magnetic field that can safely penetrate and activate the brain's inner regions has remained the Holy Grail of TMS research for some time. Creation of such a field offers the possibility of treating difficult conditions such as Parkinson's disease. Though unlikely, it might even make it possible to energize the brain's "pleasure center" directly (think "Orgasmatron," from Woody Allen's film *Sleeper*).

An interdisciplinary team at the U.S. National Institutes of Health has invented a new TMS coil configuration that is designed to generate sufficient magnetic field strength to stimulate neurons deep inside the brain mass without posing a hazard. The research group included Abraham Zangen, Roy A. Wise, Mark Hallett, Pedro C. Miranda and Yiftach Roth.

According to Zangen, now a neurobiologist at the Weizmann Institute of Science in Israel, the prototype device is designed to maximize the electric field deep in the brain by summing separate fields projected into the skull from several points around its periphery. The device also minimizes the accumulation of electrical charge on the surface of the brain, which would give rise to an electrostatic field that reduces the magnitude of the induced electric field both at the surface and inside. The unique, form-fitting shape of the base of the new stimulator positions wire coils containing several wire strips that are set tangentially to the scalp's surface. Each set of strips is connected in series and contains current flowing in the same direction. Therefore, each set generates a field that extends into the brain in a specified orientation from each location along the scalp.

The prototype apparatus underwent an initial round of clinical evaluations in the summer of 2006. Investors have recently established a company called Brainsway in Delaware to carry on the research and development effort and to commercialize the deep brain magnetic stimulator. —THE EDITORS

ELECTROMAGNETIC BRAIN-STIMULATION TECHNIQUES

Neuroscientists employ electricity and magnetism to treat brain disorders. Each method offers different degrees of targeting accuracy.

	TREATMENT USE	PULSE DELIVERY
Electroconvulsive therapy (ECT)	Depression, mania, catatonia	Skin electrodes
Transcutaneous electrical nerve stimulation (TENS)	Pain, spasticity	Skin electrodes (attached to peripheral nerves)
Vagus nerve stimulation (VNS)	Approved for epilepsy; FDA trials under way for depression and anxiety	Electrodes (attached to vagus nerve)
Deep brain stimulation (DBS)	Approved for Parkinson's disease; under investigation for pain and obsessive-compulsive disorder	Electrodes (embedded in brain regions)
Transcranial direct current stimulation (tDCS)	Under investigation for Parkinson's disease	Electric field
Transcranial magnetic stimulation (TMS)	Under investigation for depression; FDA trials under way	Magnetic field
Magnetic seizure therapy (MST)	Under investigation for depression	Magnetic field

Further complication occurs because each person's brain is wired differently, so the location for behaviors varies. If one's motor area is close to one's skull, TMS might have a large effect. In someone else, whose motor area lies deeper in, TMS may have little or no effect on movement.

To better understand the effects of rTMS on brain circuits, physicist Daryl E. Bohning and others in our group at MUSC developed the ability to perform rTMS testing in combination with a functional magnetic resonance imaging

STIMULATING THE BRAIN : 31

TARGETING ABILITY	ADVANTAGES	DISADVANTAGES
Fair	Effective for depression; side effects reduced with newer systems	Nonfocal; can lead to memory side effects; requires repeated general anesthesia
Good	Does not require surgery	Limited access to brain
Fair	Does not involve brain surgery	Effects modest (to date); unclear how to tune pulses to alter brain function
Excellent	Discrete targeting; marked effects	Potential side effects if incorrectly positioned; invasive brain surgery
Unfocused	Noninvasive	Scalp irritation; nonfocal
Excellent	Noninvasive and safe; potential for many applications	Limited to surface brain stimulation; effect on neural function still unclear
Fair	May offer better targeting and might avoid side effects of ECT	No efficacy data yet; requires repeated general anesthesia

(fMRI) scanner. Many researchers had thought that generating the powerful TMS magnetic fields within an fMRI machine was impossible or unwise. By applying rTMS within the scanner as subjects perform a task, however, one can know exactly where the stimulation is occurring and can image alterations to the neural circuit taking place because of the stimulus. Our group has shown that the brain changes that TMS causes when it makes your thumb move are very much the same as when you move your thumb in a similar pattern of your own accord. Two research groups in Germany have also succeeded in conducting rTMS studies within an fMRI scanner.

MAGNET THERAPY

In theory, TMS could be a useful therapy for any brain disorder involving dysfunctional behavior in a neural circuit. Researchers have tried employing the technique as a treatment for obsessive-compulsive disorder, schizophrenia, Parkinson's, dystonia (involuntary muscle contractions), chronic pain and epilepsy. For most of these conditions, only a few inconclusive or contradictory studies currently exist, so the jury is still out regarding the effectiveness of TMS as therapy for them.

Most of these inquiries have concentrated on relieving depression. In the mid-1990s I was among the first researchers (along with several European groups) to investigate the use of daily rTMS sessions to treat depression. Perhaps, we thought, one could accomplish what ECT does for depressed individuals with TMS while avoiding seizures. My studies (at the National Institute of Mental Health) focused on stimulating the prefrontal cortex because that region appears abnormal in many internal images of depressed patients and because it governs deeper limbic regions involved in mood and emotion regulation. Double-blind studies soon indicated a small but significant antidepressant effect. A few patients at the NIH who had not responded to any other treatments had emerged from their depression and returned home.

Since then, more than 20 randomized and controlled trials of prefrontal rTMS as a treatment for depression have been published. Most of these studies show antidepressant effects significantly greater than sham electrode application, a conclusion that has been further confirmed by subsequent meta-analyses of the results. Whereas current consensus holds that rTMS offers a statistically significant antidepressant effect, controversy continues over whether these effects are sufficient to be clinically useful.

Because no commercial industry yet exists to promote TMS as an antidepressant therapy and because most of the studies have been relatively small (with considerable variation in rTMS methods and patient selection), the use of rTMS as a treatment for depression is still considered experimental by the U.S. Food and Drug Administration. The technique has, however, already been sanctioned for use in Canada, where it is now available. A large industry-sponsored trial designed to garner FDA acceptance is being planned. Even if the approach is approved, much additional research remains to refine it.

Repetitive TMS can, it should be noted, cause seizures or epileptic convulsions in healthy subjects, depending on the intensity, frequency, duration and interval of magnetic stimuli. In the history of the technique's use, TMS has led to eight unintended seizures, but since the publication of safety guidelines several years ago, no new seizures have been reported. Some scientists are investigating the potential positive application of this result. Harold A. Sackeim and Sarah H. Lisanby of Columbia have shown that a supercharged version of TMS, which they call magnetic seizure therapy (MST), can produce beneficial seizures in depressed patients (who are first anesthetized). Unlike ECT, MST allows users to focus on the site where the seizure is triggered. Better control over the seizure should block its spread to the regions of the brain responsible for the memory loss seen with ECT. Preliminary data indicate that MST has fewer cognitive side effects than traditional ECT techniques. More needs to be done to determine whether the MST really works and for which disorders it might be beneficial.

The technology of TMS is evolving as well. Our group at MUSC, for instance, has recently developed a portable TMS machine—an advance that may someday translate into the fatigue-fighting flight helmets depicted earlier. Extensive development is also proceeding on new designs and prototypes for coils that can stimulate more deeply inside the brain, that can be focused more finely or that operate in coordinated arrays. Most of our actions and thoughts arise not from activity in a single brain region but rather through the coordinated firing of many brain regions. If one could make several TMS coils, distributed over various key regions and fired in a coordinated way, new vistas might open up for TMS as a neuroscience tool and treatment.

After more than a decade of experimentation, TMS is still not FDA-approved to alleviate any disorder. Nevertheless, interest remains high among researchers who continue to believe in the intuitive aptness of using safe magnetic fields to turn specific brain regions on and off. If TMS proves itself, it could even lend some credence to the folk wisdom that humans use only a small portion of their brains.

MORE TO EXPLORE

Barker, Anthony T., Reza Jalinous, and Ian L. Freeston. 1985. Non-invasive magnetic stimulation of the human motor cortex. *Lancet* 1:1106–1107.

Bohning, Daryl E. 2000. Introduction and overview of TMS physics. In *Transcranial magnetic stimulation in neuropsychiatry*, ed. Mark S. George and Robert H. Belmaker. American Psychiatric Press.

Bohning, D. E., A. Shastri, Z. Nahas, J. P. Lorberbaum, S. W. Andersen, W. R. Dannels, E. U. Haxthausen, D. J. Vincent, and M. S. George. 1998. Echoplanar BOLD fMRI of brain activation induced by concurrent transcranial magnetic stimulation (TMS). *Investigative Radiology* 33 (6): 336–340.

Boroojerdi, B., M. Phipps, L. Kopylev, C. M. Wharton, L. G. Cohen, and J. Grafman. 2001. Enhancing analogic reasoning with rTMS over the left prefrontal cortex. *Neurology* 56 (4): 526–528.

George, M. S., and R. H. Belmaker. 2000. *TMS in neuropsychiatry.* American Psychiatric Press.

George, M. S., E. M. Wassermann, W. Williams, A. Callahan, T. A. Ketter, P. Basser, M. Hallett, and R. M. Post. 1995. Daily repetitive transcranial magnetic stimulation (rTMS) improves mood in depression. *NeuroReport,* no. 14 (October 2): 1853–1856.

For more information on transcranial magnetic stimulation, visit pni.unibe.ch/TMS.htm.

MARK S. GEORGE is a practicing psychiatrist and neurologist as well as a research neuroscientist. George first studied the relation between mind and brain as an undergraduate philosophy student at Davidson College. His fascination with the human brain continued throughout medical school and dual residencies at the Medical University of South Carolina (MUSC). George investigated and developed new brain imaging and brain-stimulation techniques during fellowships at the Institute of Neurology in London and the National Institutes of Health. He returned to MUSC eight years ago to run laboratories devoted to brain imaging and brain-stimulation research. Originally published in *Scientific American,* Vol. 289, No. 3, September 2003.

3

Freud Returns

Neuroscientists are finding that their biological descriptions of the brain may fit
together best when integrated by psychological theories that Freud sketched a
century ago

MARK SOLMS

The founder of psychoanalysis was born 150 years ago, and in 2006 his theories
are enjoying a rebirth. New life indeed, because not too long ago his ideas were
considered dead.

For the first half of the 1900s, Sigmund Freud's explanations dominated
views of how the human mind works. His basic proposition was that our
motivations remain largely hidden in our unconscious minds. Moreover, they
are actively withheld from consciousness by a repressive force. The executive
apparatus of the mind (the ego) rejects any unconscious drives (the id) that
might prompt behavior that would be incompatible with our civilized conception of ourselves. This repression is necessary because the drives express themselves in unconstrained passions, childish fantasies, and sexual and aggressive
urges.

Mental illness, Freud said until his death in 1939, results when repression
fails. Phobias, panic attacks and obsessions are caused by intrusions of the hidden drives into voluntary behavior. The aim of psychotherapy, then, was to trace
neurotic symptoms back to their unconscious roots and expose these roots to
mature, rational judgment, thereby depriving them of their compulsive power.

As mind and brain research grew more sophisticated from the 1950s on-
ward, however, it became apparent to specialists that the evidence Freud had
provided for his theories was rather tenuous. His principal method of investigation was not controlled experimentation but simple observations of patients
in clinical settings, interwoven with theoretical inferences. Drug treatments

gained ground, and biological approaches to mental illness gradually overshadowed psychoanalysis. Had Freud been alive, he might even have welcomed this turn of events. A highly regarded neuroscientist in his day, he frequently made remarks such as "the deficiencies in our description would presumably vanish if we were already in a position to replace the psychological terms by physiological and chemical ones." But Freud did not have the science or technology to know how the brain of a normal or neurotic personality was organized.

By the 1980s the notions of ego and id were considered hopelessly antiquated, even in some psychoanalytic circles. Freud was history. In the new psychology, the updated thinking went, depressed people do not feel so wretched because something has undermined their earliest attachments in infancy—rather their brain chemicals are unbalanced. Psychopharmacology, however, did not deliver an alternative grand theory of personality, emotion and motivation—a new conception of "what makes us tick." Without this model, neuroscientists focused their work narrowly and left the big picture alone.

Today that picture is coming back into focus, and the surprise is this: it is not unlike the one that Freud outlined a century ago. We are still far from a consensus, but an increasing number of diverse neuroscientists are reaching the same conclusion drawn by Eric R. Kandel of Columbia University, the 2000 Nobel Laureate in Physiology or Medicine: that psychoanalysis is "still the most coherent and intellectually satisfying view of the mind."

Freud is back, and not just in theory. Interdisciplinary work groups uniting the previously divided and often antagonistic fields of neuroscience and psychoanalysis have been formed in almost every major city of the world. These networks, in turn, have come together as the International Neuro-Psychoanalysis Society, which organizes an annual congress and publishes the successful journal *Neuro-Psychoanalysis*. Testament to the renewed respect for Freud's ideas is the journal's editorial advisory board, populated by a who's who of experts in contemporary behavioral neuroscience, including Antonio R. Damasio, Kandel, Joseph E. LeDoux, Benjamin Libet, Jaak Panksepp, Vilayanur S. Ramachandran, Daniel L. Schacter and Wolf Singer.

Together these researchers are forging what Kandel calls a "new intellectual framework for psychiatry." Within this framework, it appears that Freud's broad brushstroke organization of the mind is destined to play a role similar to the one Darwin's theory of evolution served for molecular genetics—a template on which emerging details can be coherently arranged. At the same time, neuroscientists are uncovering proof for some of Freud's theories and are teasing out the mechanisms behind the mental processes he described.

UNCONSCIOUS MOTIVATION

When Freud introduced the central notion that most mental processes that determine our everyday thoughts, feelings and volitions occur unconsciously, his contemporaries rejected it as impossible. But today's findings are confirming the existence and pivotal role of unconscious mental processing. For example, the behavior of patients who are unable to consciously remember events that occurred after damage to certain memory-encoding structures of their brains is clearly influenced by the "forgotten" events.

Cognitive neuroscientists make sense of such cases by delineating different memory systems that process information "explicitly" (consciously) and "implicitly" (unconsciously). Freud split memory along just these lines.

Neuroscientists also have identified unconscious memory systems that mediate emotional learning. In 1996 at New York University, LeDoux demonstrated the existence under the conscious cortex of a neuronal pathway that connects perceptual information with the primitive brain structures responsible for generating fear responses.

Because this pathway bypasses the hippocampus—which generates conscious memories—current events routinely trigger unconscious remembrances of emotionally important past events, causing conscious feelings that seem irrational, such as "Men with beards make me uneasy."

Neuroscience has shown that the major brain structures essential for forming conscious (explicit) memories are not functional during the first two years of life, providing an elegant explanation of what Freud called infantile amnesia. As Freud surmised, it is not that we forget our earliest memories; we simply cannot recall them to consciousness. But this inability does not preclude them from affecting adult feelings and behavior. One would be hard-pressed to find a developmental neurobiologist who does not agree that early experiences, especially between mother and infant, influence the pattern of brain connections in ways that fundamentally shape our future personality and mental health. Yet none of these experiences can be consciously remembered. It is becoming increasingly clear that a good deal of our mental activity is unconsciously motivated.

REPRESSION VINDICATED

Even if we are mostly driven by unconscious thoughts, this does not prove anything about Freud's claim that we actively repress unpalatable information. But case studies supporting that notion are beginning to accumulate. The most famous one comes from a 1994 study of "anosognosic" patients by Ramachandran, a behavioral neurologist at the University of California, San Diego. Damage to

MIND AND MATTER

Freud drew his final model of the mind in 1933 (a; color has been added). Dotted lines represented the threshold between unconscious and conscious processing. The superego repressed instinctual drives (the id), preventing them from disrupting rational thought. Most rational (ego) processes were automatic and unconscious, too, so only a small part of the ego (bulb at top) was left to manage conscious experience, which was closely tied to perception. The superego mediated the ongoing struggle between the ego and id for dominance. Recent neurological mapping (b) generally correlates to Freud's conception. The core brain stem and limbic system—responsible for instincts and drives—roughly correspond to Freud's id. The ventral frontal region, which controls selective inhibition, the dorsal frontal region, which controls self-conscious thought, and the posterior cortex, which represents the outside world, amount to the ego and the superego.

(a)

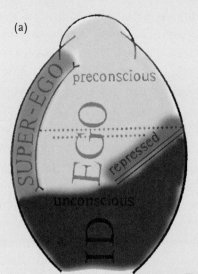

(b)

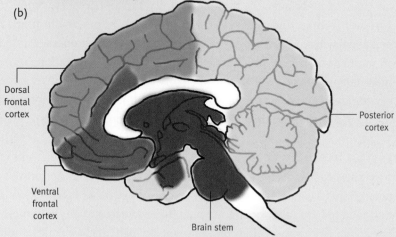

Dorsal frontal cortex

Posterior cortex

Ventral frontal cortex

Brain stem

the right parietal region of these people's brains makes them unaware of gross physical defects, such as paralysis of a limb. After artificially activating the right hemisphere of one such patient, Ramachandran observed that she suddenly became aware that her left arm was paralyzed—and that it had been paralyzed continuously since she had suffered a stroke eight days before. This showed that she was capable of recognizing her deficits and that she had unconsciously registered these deficits for the previous eight days, despite her conscious denials during that time that there was any problem.

Significantly, after the effects of the stimulation wore off, the woman not only reverted to the belief that her arm was normal, she also forgot the part of the interview in which she had acknowledged that the arm was paralyzed, even though she remembered every other detail about the interview. Ramachandran concluded: "The remarkable theoretical implication of these observations is that memories can indeed be selectively repressed. . . . Seeing [this patient] convinced me, for the first time, of the reality of the repression phenomena that form the cornerstone of classical psychoanalytical theory."

Like "split-brain" patients, whose hemispheres become unlinked—a situation made famous in studies by Nobel laureate Roger W. Sperry of the California Institute of Technology in the 1960s and 1970s—anosognosic patients typically rationalize away unwelcome facts, giving plausible but invented explanations of their unconsciously motivated actions. In this way, Ramachandran says, the left hemisphere manifestly employs Freudian "mechanisms of defense."

Analogous phenomena have now been demonstrated in people with intact brains, too. As neuropsychologist Martin A. Conway of Durham University in England pointed out in a 2001 commentary in *Nature,* if significant repression effects can be generated in average people in an innocuous laboratory setting, then far greater effects are likely in real-life traumatic situations.

THE PLEASURE PRINCIPLE

Freud went even further, though. He said that not only is much of our mental life unconscious and withheld but the repressed part of the unconscious mind operates according to a different principle than the "reality principle" that governs the conscious ego. This type of unconscious thinking is "wishful"—and it blithely disregards the rules of logic and the arrow of time.

If Freud was right, then damage to the inhibitory structures of the brain (the seat of the "repressing" ego) should release wishful, irrational modes of mental functioning.

This is precisely what has been observed in patients with damage to the frontal limbic region, which controls critical aspects of self-awareness. Subjects display a striking syndrome known as Korsakoff's psychosis: they are unaware that they are amnesic and therefore fill the gaps in their memory with fabricated stories known as confabulations.

Durham neuropsychologist Aikaterini Fotopoulou studied a patient of this type in my laboratory. The man failed to recall, in each 50-minute session held in my office on 12 consecutive days, that he had ever met me before or that he had undergone an operation to remove a tumor in his frontal lobe that caused his amnesia. As far as he was concerned, there was nothing wrong with him. When asked about the scar on his head, he confabulated wholly implausible explanations: he had undergone dental surgery or a coronary bypass operation.

In reality, he had indeed experienced these procedures—years before—and unlike his brain operation, they had successful outcomes.

Similarly, when asked who I was and what he was doing in my lab, he variously said that I was a colleague, a drinking partner, a client consulting him about his area of professional expertise, a teammate in a sport that he had not participated in since he was in college decades earlier, or a mechanic repairing one of his numerous sports cars (which he did not possess). His behavior was consistent with these false beliefs, too: he would look around the room for his beer or out the window for his car.

What strikes the casual observer is the wishful quality of these false notions, an impression that Fotopoulou confirmed objectively through quantitative analysis of a consecutive series of 155 of his confabulations. The patient's false beliefs were not random noise—they were generated by the "pleasure principle" that Freud maintained was central to unconscious thought. The man simply recast reality as he wanted it to be. Similar observations have been reported by others, such as Conway and Oliver Turnbull of the University of Wales. These investigators are cognitive neuroscientists, not psychoanalysts, yet they interpret their findings in Freudian terms. They claim in essence that damage to the frontal limbic region that produces confabulations impairs cognitive-control mechanisms that underpin normal reality monitoring and releases from inhibition the implicit wishful influences on perception, memory and judgment.

ANIMAL WITHIN

Freud argued that the pleasure principle gave expression to primitive, animal drives. To his Victorian contemporaries, the implication that human behavior was at bottom governed by urges that served no higher purpose than carnal self-fulfillment was downright scandalous. The moral outrage waned during subse-

Freud sketched a neuronal mechanism for repression in 1895, top left, as part of his hope that biological explanations of the mind would one day replace psychological ones. In his scheme, an unpleasant memory would normally be activated by a stimulus ("Qn," far left) heading from neuron "a" toward neuron "b" (bottom). But neuron "alpha" (to right of "a") could divert the signal and thus prevent the activation if other neurons (top right) exerted a "repressing" influence. Note that Freud drew gaps between neurons that he predicted would act as "contact barriers." Two years later English physiologist Charles Sherrington discovered such gaps and named them synapses.

quent decades, but Freud's concept of man-as-animal was pretty much sidelined by cognitive scientists.

Now it has returned. Neuroscientists such as Donald W. Pfaff of the Rockefeller University and Panksepp of Bowling Green State University believe that the instinctual mechanisms that govern human motivation are even more primitive than Freud imagined. We share basic emotional-control systems with our primate relatives and with all mammals. At the deep level of mental organization that Freud called the id, the functional anatomy and chemistry of our brains are not much different from those of our favorite barnyard animals and household pets.

Modern neuroscientists do not accept Freud's classification of human instinctual life as a simple dichotomy between sexuality and aggression, however. Instead, through studies of lesions and the effects of drugs and artificial stimulation on the brain, they have identified at least four basic mammalian instinctual circuits, some of which overlap.

They are the "seeking" or "reward" system (which motivates the pursuit of pleasure); the "anger-rage" system (which governs angry aggression but not predatory aggression); the "fear-anxiety" system; and the "panic" system (which includes complex instincts such as those that govern social bonding). Whether other instinctual forces exist, such as a rough-and-tumble "play" system, is also being investigated. All these systems are modulated by specific neurotransmitters, chemicals that carry messages between neurons.

The seeking system, regulated by dopamine, bears a remarkable resemblance to the Freudian "libido." According to Freud, the libidinal or sexual drive is a pleasure-seeking system that energizes most of our goal-directed behavior. Modern research shows that its neural equivalent is heavily implicated in almost all forms of craving and addiction. It is interesting to note that Freud's early experiments with cocaine—mainly on himself—convinced him that the libido must have a specific neurochemical foundation. Unlike his successors, Freud saw no reason for antagonism between psychoanalysis and psychopharmacology. He enthusiastically anticipated the day when "id energies" would be controlled directly by "particular chemical substances." Today treatments that integrate psychotherapy with psychoactive medications are widely recognized as the best approach for many disorders. And brain imaging shows that some talk therapy affects the brain in similar ways to such drugs.

DREAMS HAVE MEANING

Freud's ideas are also reawakening in sleep and dream science. His dream theory—that nighttime visions are partial glimpses of unconscious wishes—was discredited when rapid eye movement (REM) sleep and its strong correlation with dreaming were discovered in the 1950s. Freud's view appeared to lose all credibility when investigators in the 1970s showed that the dream cycle was regulated by the pervasive brain chemical acetylcholine. REM sleep occurred automatically, every 90 minutes or so, and was driven by brain chemicals and structures that had nothing to do with emotion or motivation. This discovery implied that dreams had no meaning; they were simply stories concocted by the higher brain to try to reflect the random cortical activity caused by REM.

But more recent work has revealed that dreaming and REM sleep are dissociable states, controlled by distinct, though interactive, mechanisms. Dreaming turns out to be generated by the forebrain's instinctual-motivational circuitry. This discovery has given rise to a host of theories about the dreaming brain, many strongly reminiscent of Freud's. Most intriguing is the observation that others and I have made that dreaming stops completely when certain fibers deep in the frontal lobe have been severed—a symptom that coincides with a general reduction in motivated behavior. The lesion is the same as the damage that was deliberately produced in prefrontal leukotomy, an outmoded surgical procedure once used to control hallucinations and delusions. This operation was replaced in the 1960s by drugs that dampen dopamine's activity in the same brain systems. The seeking system, then, might be the primary generator of dreams. This possibility has become a major focus of current research.

If the hypothesis is confirmed, then the wish-fulfillment theory of dreams could once again set the agenda for sleep research. But even if other interpretations of the new neurological data prevail, all of them demonstrate that "psychological" conceptualizations of dreaming are scientifically respectable again. Few neuroscientists still claim—as they once did with impunity—that dream content has no primary emotional mechanism.

FINISHING THE JOB

Not everyone is enthusiastic about the reappearance of Freudian concepts in mental science. It is not easy for the older generation of psychoanalysts, for example, to accept that their junior colleagues now can and must subject conventional wisdom to an entirely new level of biological scrutiny. But an encouraging number of elders on both sides of the Atlantic are at least committed to keeping an open mind.

For older neuroscientists, resistance to the return of psychoanalytical ideas comes from the specter of the seemingly indestructible edifice of Freudian theory in the early years of their careers. They cannot acknowledge even partial

FREUD RETURNS? LIKE A BAD DREAM
Counterpoint
BY J. ALLAN HOBSON

Sigmund Freud's views on the meaning of dreams formed the core of his theory of mental functioning. Mark Solms and others assert that modern science is now validating Freud's conception of the mind. But similar scientific investigations show that major aspects of Freud's thinking are probably erroneous.

For Freud, the bizarre nature of dreams resulted from an elaborate effort of the mind to conceal, by symbolic disguise and censorship, the unacceptable instinctual wishes welling up from the unconscious when the ego relaxes its prohibition of the id in sleep. But most neurobiological evidence supports the alternative view that dream bizarreness stems from normal changes in brain state. Chemical mechanisms in the brain stem, which shift the activation of various regions of the cortex, generate these changes. Many studies have indicated that the chemical changes determine the quality and quantity of dream visions, emotions and thoughts. Freud's disguise-and-censorship notion must be discarded; no one believes that the ego-id struggle, if it exists, controls brain chemistry. Most psychoanalysts no longer hold that the disguise-censorship theory is valid.

Without disguise and censorship, what is left of Freud's dream theory? Not much—only that instinctual drives could impel dream formation. Evidence does indicate that activating the parts of the limbic system that produce anxiety, anger and elation shapes dreams. But these influences are not "wishes." Dream analyses show that the emotions in dreams are as often negative as they are positive, which would mean that half our "wishes" for ourselves are negative. And as all dreamers know, the emotions in dreams are hardly disguised. They enter into dream plots clearly, frequently bringing unpleasant effects such as nightmares. Freud was never able to account for why so many dream emotions are negative.

Another pillar of Freud's model is that because the true meaning of dreams is hidden, the emotions they reflect can be revealed only through his wild-goose-chase method of free association, in which the subject relates anything and everything that comes to mind in hopes of stumbling across a crucial connection. But this effort is unnecessary, because no such concealment occurs. In dreams, what you see is what you get. Dream content is emotionally salient on its face, and the close attention of dreamers and their therapists is all that is needed to see the feelings they represent. Solms and other Freudians intimate that ascribing dreams to brain chemistry is the same as saying that dreams have no emotional messages. But the statements are not equivalent. The chemical activation-synthesis theory of dreaming, put forth by Robert W. McCarley of Harvard Medical School and me in 1977, maintained only that the psychoanalytic explanation of dream bizarreness as concealed meaning was wrong. We have always argued that dreams are emotionally salient and meaningful. And what about REM sleep? New studies reveal that dreams can occur during non-REM sleep, but nothing in the chemical activation model precludes this case; the frequency of dreams is simply exponentially higher during REM sleep. Psychoanalysis is in big trouble, and no amount of neurobiological tinkering can fix it. So radical an overhaul is necessary that many neuroscientists would prefer to start over and create a neurocognitive model of the mind. Psychoanalytic theory is indeed comprehensive, but if it is terribly in error, then its comprehensiveness is hardly a virtue. The scientists who share this view stump for more biologically-based models of dreams, of mental illness, and of normal conscious experience than those offered by psychoanalysis.

confirmation of Freud's fundamental insights; they demand a complete purge. In the words of J. Allan Hobson, a renowned sleep researcher and Harvard Medical School psychiatrist, the renewed interest in Freud is little more than unhelpful "retrofitting" of modern data into an antiquated theoretical framework. But as Fred Guterl wrote in a 2002 interview with Panksepp in *Newsweek* magazine, for neuroscientists who are enthusiastic about the reconciliation of neurology and psychiatry, "it is not a matter of proving Freud wrong or right, but of finishing the job."

If that job can be finished—if Kandel's "new intellectual framework for psychiatry" can be established—then the time will pass when people with emotional difficulties have to choose between the talk therapy of psychoanalysis, which may be out of touch with modern evidence-based medicine, and the drugs prescribed by psychopharmacology, which may lack regard for the relation between the brain chemistries it manipulates and the complex real-life trajectories that culminate in emotional distress. The psychiatry of tomorrow promises to provide patients with help that is grounded in a deeply integrated understanding of how the human mind operates. Whatever undreamed-of therapies the future might bring, patients can only benefit from better knowledge of how the brain really works. As modern neuroscientists tackle once more the profound questions of human psychology that so preoccupied Freud, it is gratifying to find that we can build on the foundations he laid, instead of having to start all over again. Even as we identify the weak points in Freud's far-reaching theories, and thereby correct, revise and supplement his work, we are excited to have the privilege of finishing the job.

MORE TO EXPLORE

Solms, Mark. 2000. Freudian dream theory today. *Psychologist* 13 (12): 618–619.
Solms, Mark, and Oliver Turnbull. *The brain and the inner world.* Other Press.
International Neuro–Psychoanalysis Society. *Neuro-Psychoanalysis.*
  www.neuropsa.org.uk

MARK SOLMS holds the chair in neuropsychology at the University of Cape Town in South Africa and an honorary lectureship in neurosurgery at St. Bartholomew's and the Royal London School of Medicine and Dentistry. He is also director of the Arnold Pfeffer Center for Neuro-Psychoanalysis, a consultant neuropsychologist to the Anna Freud Center in London and a very frequent flier. He thanks Oliver Turnbull, a senior lecturer at the University of Wales, for assisting with this article. Originally published in *Scientific American Mind*, Vol. 17, No. 2, April/May 2006.

4

The Neurobiology of the Self

Biologists are beginning to tease out how the brain gives rise to a constant
sense of being oneself

BY CARL ZIMMER

The most obvious thing about yourself is your self. "You look down at your body
and know it's yours," says Todd Heatherton, a psychologist at Dartmouth University. "You know it's your hand you're controlling when you reach out. When
you have memories, you know that they are yours and not someone else's. When
you wake up in the morning, you don't have to interrogate yourself for a long
time about who you are."

The self may be obvious, but it is also an enigma. Heatherton himself shied
away from direct study of it for years, even though he had been exploring self-
control, self-esteem and other related issues since graduate school. "My interests
were all around the self but not around the philosophical issue of what is the self,"
he explains. "I avoided speculations about what it means. Or I tried to, anyway."

Things have changed. Today Heatherton, along with a growing number of
other scientists, is tackling this question head-on, seeking to figure out how the
self emerges from the brain. In the past few years, they have begun to identify
certain brain activities that may be essential for producing different aspects of
self-awareness. They are now trying to determine how these activities give rise
to the unified feeling we each have of being a single entity. This research is
yielding clues to how the self may have evolved in our hominid ancestors. It may
even help scientists treat Alzheimer's disease and other disorders that erode the
knowledge of self and, in some cases, destroy it altogether.

THE SELF IS SPECIAL

American psychologist William James launched the modern study of this area
in 1890 with his landmark book, *The Principles of Psychology*. "Let us begin with
the Self in its widest acceptation, and follow it up to its most delicate and subtle

OVERVIEW/MY BRAIN AND ME
- Increasing numbers of neurobiologists are exploring how the brain manages to form and maintain a sense of self.
- Several brain regions have been found to respond differently to information relating to the self than they do to information relating to others, even to very familiar others. For instance, such regions may be more active when people think about their own attributes than when they think about the characteristics of other individuals. These regions could be part of a self-network.
- For some, the goal of this research is to better understand, and to find new therapies for, dementia.

form," he proposed. James argued that although the self might feel like a unitary thing, it has many facets—from awareness of one's own body to memories of oneself to the sense of where one fits into society. But James confessed to being baffled as to how the brain produced these self-related thoughts and wove them into a single ego.

Since then, scientists have found some telling clues through psychological experiments. Researchers interested in memories of the self, for instance, have asked volunteers questions about themselves, as well as about other people. Later the researchers gave the volunteers a pop quiz to see how well they remembered the questions. People consistently did a better job of remembering questions about themselves than about others. "When we tag things as relevant to the self, we remember them better," Heatherton says.

Some psychologists argued that these results simply meant that we are more familiar to ourselves than other people are to us. Some concluded instead that the self is special; the brain uses a different, more efficient system to process information about it. But psychological tests could not pick a winner from these competing explanations, because in many cases the hypotheses made the same predictions about experimental outcomes.

Further clues have emerged from injuries that affect some of the brain regions involved in the self. Perhaps the most famous case was that of Phineas Gage, a 19th-century railroad construction foreman who was standing in the wrong place at the wrong time when a dynamite blast sent a tamping iron through the air. It passed right through Gage's head, and yet, astonishingly, Gage survived.

Gage's friends, though, noticed something had changed. Before the accident, he had been considered an efficient worker and a shrewd businessman. Afterward he became profane, showed little respect for others and had a hard time settling on plans for the future. His friends said he was "no longer Gage."

Cases such as Gage's showed that the self is not the same as consciousness. People can have an impaired sense of themselves without being unconscious. Brain injuries have also revealed that the self is constructed in a complicated way. In 2002, for example, Stan B. Klein of the University of California at Santa Barbara and his colleagues reported on an amnesiac known as D.B. The man was 75 years old when he suffered brain damage from a heart attack and lost the ability to recall anything he had done or experienced before it. Klein tested D.B.'s self-knowledge by giving him a list of 60 traits and asking him whether they applied to him somewhat, quite a bit, definitely, or not at all. Then Klein gave the same questionnaire to D.B.'s daughter and asked her to use it to describe her father. D.B.'s choices significantly correlated with his daughter's. Somehow D.B. had retained an awareness of himself without any access to memories of who he was.

CLUES FROM HEALTHY BRAINS

In recent years, scientists have moved beyond injured brains to healthy ones, thanks to advances in brain imaging. At University College London, researchers have been using brain scans to decipher how we become aware of our own bodies. "This is the very basic, low-level first point of the self," UCL's Sarah-Jayne Blakemore says.

When our brains issue a command to move a part of our bodies, two signals are sent. One goes to the brain regions that control the particular parts of the body that need to move, and another goes to regions that monitor the movements. "I like to think of it as a 'cc' of an e-mail," Blakemore observes. "It's all the same information sent to a different place."

Our brains then use this copy to predict what kind of sensation the action will produce. A flick of an eye will make objects appear to move across our field of vision. Speaking will make us hear our own voice. Reaching for a doorknob will make us feel the cold touch of brass. If the actual sensation we receive does not closely match our prediction, our brains become aware of the difference. The mismatch may make us pay more attention to what we are doing or prompt us to adjust our actions to get the results we want.

But if the sensation does not match our predictions at all, our brains interpret it as being caused by something other than ourselves. Blakemore and her colleagues documented this shift by scanning the brains of subjects they had hypnotized. When the researchers told the subjects their arms were being lifted by a rope and pulley, the subjects lifted their arms. But their brains responded as if someone else were lifting their arms, not themselves.

A similar lack of self-awareness may underlie certain symptoms of schizo-phrenia. Some schizophrenics become convinced that they cannot control their own bodies. "They reach over to grab a glass, and their movement is totally nor-mal. But they say, 'It wasn't me. That machine over there controlled me and made me do it,'" Blakemore explains.

Studies on schizophrenics suggest that bad predictions of their own actions may be the source of these delusions. Because their sensations do not match their predictions, it feels as if something else is responsible. Bad predictions may also create the auditory hallucinations that some schizophrenics experience. Unable to predict the sound of their inner voice, they think it belongs to someone else.

One reason the sense of self can be so fragile may be that the human mind is continually trying to get inside the minds of other people. Scientists have discov-ered that so-called mirror neurons mimic the experiences of others. The sight of someone being painfully poked, for example, stimulates neurons in the pain region of our own brains. Blakemore and her colleagues have found that even seeing someone touched can activate mirror neurons.

They recently showed a group of volunteers videos of other people being touched on the left or right side of the face or neck. The videos elicited the same responses in some areas of the volunteers' brains as occurred when the volun-teers were touched on the corresponding parts of their own bodies. Blakemore was inspired to carry out the study when she met a 41-year-old woman, known as C., who took this empathy to a surprising extreme. The sight of someone being touched made C. feel as if someone were touching her in the same place on her own body. "She thought everyone had that experience," Blakemore remarks.

Blakemore scanned the woman's brain and compared its responses to those of normal volunteers. C.'s touch-sensitive regions reacted more strongly to the sight of someone else being touched than those regions did in the normal sub-jects. In addition, a site called the anterior insula (located on the brain's sur-face not far from the ear) became active in C. but not in the normal volunteers. Blakemore finds it telling that the anterior insula has also displayed activity in brain scans of people who are shown pictures of their own faces or who are identifying their own memories. It is possible that the anterior insula helps to designate some information as relating to ourselves instead of to other people. In the case of C., it simply assigns information incorrectly.

Brain scans have also shed light on other aspects of the self. Heatherton and his colleagues at Dartmouth have been using the technology to probe the mys-tery of why people remember information about themselves better than details about other people. They imaged the brains of volunteers who viewed a series of adjectives. In some cases, the researchers asked the subjects whether a word applied to the subjects themselves. In others, they asked if a word applied to George W. Bush. In still other cases, they asked simply whether the word was shown in uppercase letters.

The researchers then compared the patterns of brain activity triggered by each kind of question. They found that questions about the self activated some regions of the brain that questions about someone else did not. Their results bolstered the "self is special" hypothesis over the "self is familiar" view.

A COMMON DENOMINATOR

One region that Heatherton's team found to be important to thinking about one-self was the medial prefrontal cortex, a patch of neurons located in the cleft be-tween the hemispheres of the brain, directly behind the eyes. The same region also has drawn attention in studies on the self carried out by other laboratories. Heatherton is now trying to figure out what role it serves.

"It's ludicrous to think that there's any spot in the brain that's 'the self,'" he says. Instead he suspects that the area may bind together all the perceptions and memories that help to produce a sense of self, creating a unitary feeling of who we are. "Maybe it's something that brings information together in a meaningful way," Heatherton notes.

If he is right, the medial prefrontal cortex may play the same role for the self as the hippocampus plays in memory. The hippocampus is essential for form-ing new memories, but people can still retain old memories even after it is in-jured. Rather than storing information on its own, the hippocampus is believed to create memories by linking together far-flung parts of the brain.

The medial prefrontal cortex could be continuously stitching together a sense of who we are. Debra A. Gusnard of Washington University and her co-workers have investigated what occurs in the brain when it is at rest—that is, not en-gaged in any particular task.

It turns out that the medial prefrontal cortex becomes more active at rest than during many kinds of thinking.

"Most of the time we daydream—we think about something that happened to us or what we think about other people. All this involves self-reflection," Heath-erton says.

Other scientists are investigating the brain networks that may be organized by the medial prefrontal cortex. Matthew Lieberman of the University of Cali-fornia at Los Angeles has been using brain scans to solve the mystery of D.B., the man who knew himself even though he had amnesia. Lieberman and his colleagues scanned the brains of two sets of volunteers: soccer players and im-provisational actors. The researchers then wrote up two lists of words, each of which was relevant to one of the groups. (Soccer players: athletic, strong, swift; actors: performer, dramatic, and so on.) They also composed a third list of words that did not apply specifically to either (messy and reliable, for example). Then they showed their subjects the words and asked them to decide whether each one applied to themselves or not.

The volunteers' brains varied in their responses to the different words.

JUST ANOTHER PRETTY FACE?

As Carl Zimmer notes in the accompanying article, investigators disagree over whether the brain treats the self as special—processing information about the self differently from information about other aspects of

J.W.   Gazzaniga

life. Some argue that parts of our brain that change their activity when we think about ourselves do so simply because we are familiar with ourselves, not specifically because the self is involved; anything else that was familiar would evoke the same response.

In one study addressing this question, researchers photographed a man referred to as J.W., whose right and left cerebral hemispheres operated independently as a result of surgery that had severed the connections (to treat intractable epilepsy). They also photographed someone very familiar to the man—Michael Gazzaniga, a well-known brain researcher who had spent a lot of time with J.W. Next they created a series of images in which J.W.'s face morphed into Gazzaniga's (below) and displayed them in random order. For each image, they had J.W. answer the question "Is it me?" Then they repeated the process, having him answer, "Is it Mike?" They also performed the test with the faces of others well known to J.W.

They found that J.W.'s right hemisphere was more active when he recognized familiar others, but his left hemisphere was most active when he saw himself in the photographs. These findings lend support to the self-is-special hypothesis. The issue, though, is far from solved: both camps have evidence in their favor.—RICKI RUSTING, MANAGING EDITOR, SCIENTIFIC AMERICAN

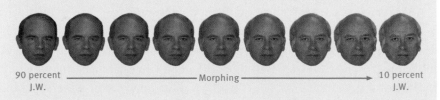

90 percent J.W. ———————————— Morphing ——————————→ 10 percent J.W.

COMPONENTS OF A SELF-NETWORK

The brain regions highlighted below are among those that have been implicated, at least by some studies, as participating in processing or retrieving information specifically related to the self or in maintaining a cohesive sense of self across all situations. For clarity, the view below omits the left hemisphere, except for its anterior insula region.

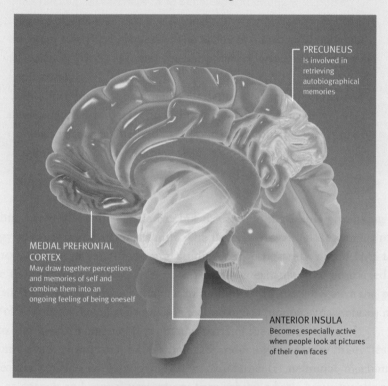

PRECUNEUS
Is involved in retrieving autobiographical memories

MEDIAL PREFRONTAL CORTEX
May draw together perceptions and memories of self and combine them into an ongoing feeling of being oneself

ANTERIOR INSULA
Becomes especially active when people look at pictures of their own faces

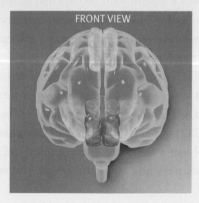

FRONT VIEW

Soccer-related words tended to increase activity in a distinctive network in the brains of soccer players, the same one that became more active in response to actor-related words in actors. When they were shown words related to the other group, a different network became more active. Lieberman refers to these two networks as the reflective system (or C system) and the reflexive system (or X system).

The C system taps into the hippocampus and other parts of the brain already known to retrieve memories. It also includes regions that can consciously hold pieces of information in mind. When we are in new circumstances, our sense of our self depends on thinking explicitly about our experiences.

But Lieberman argues that over time, the X system takes over. Instead of memories, the X system encodes intuitions, tapping into regions that produce quick emotional responses based not on explicit reasoning but on statistical associations. The X system is slow to form its self-knowledge, because it needs many experiences to form these associations. But once it takes shape, it becomes very powerful. Soccer players know whether they are athletic, strong or swift without having to consult their memories. Those qualities are intimately wrapped up with who they are. On the other hand, they do not have the same gut instinct about whether they are dramatic, and in these cases they must think explicitly about their experiences. Lieberman's results may solve the mystery of D.B. 's paradoxical self-knowledge. It is conceivable that his brain damage wiped out his reflective system but not his reflexive system.

Although the neuroscience of the self is now something of a boom industry, it has its critics. "A lot of these studies aren't constrained, so they don't say anything," says Martha Farah, a cognitive neuroscientist at the University of Pennsylvania. The experiments, she argues, have not been designed carefully enough to eliminate other explanations—for example, that we use certain brain regions to think about any person, including ourselves. "I don't think there's any 'there' there," she says.

Heatherton and other scientists involved in this research think that Farah is being too tough on a young field. Still, they agree that they have yet to figure out much about the self-network and how it functions.

THE EVOLVING SELF

Uncovering this network may allow scientists to understand how our sense of self evolved. The primate ancestors of humans probably had the basic bodily self-awareness that is studied by Blakemore and her associates. (Studies on monkeys suggest that they make predictions about their own actions.) But humans have evolved a sense of self that is unparalleled in its complexity. It may be significant that the medial prefrontal cortex is "one of the most distinctly human brain regions," according to Lieberman. Not only is it larger in humans than in nonhuman primates, but it also has a greater concentration of uniquely shaped neurons called spindle cells. Scientists do not yet know what these neurons do but suspect that they play an important role in processing information. "It does seem like there's something special there," he comments.

Heatherton thinks that the human self-network may have evolved in response to the complex social life of our ancestors. For millions of years hominids lived in small bands, cooperating to find food and sharing what they found. "The only way that works is through self-control," he says. "You have to have cooperation, and you have to have trust." And these kinds of behaviors, he argues, require a sophisticated awareness of oneself.

If the full-fledged human self were a product of hominid society, that link would explain why there are so many tantalizing overlaps between how we think about ourselves and how we think about others. This overlap is not limited to the physical empathy that Blakemore studies. Humans are also uniquely skilled at inferring the intentions and thoughts of other members of their species. Scientists have scanned people engaged in using this so-called theory of mind, and some of the regions of the brain that become active are part of the network used in thinking about oneself (including the medial prefrontal cortex). "Understanding ourselves and having a theory of mind are closely related," Heatherton says. "You need both to be a functioning human being."

The self requires time to develop fully. Psychologists have long recognized that it takes a while for children to acquire a stable sense of who they are. "They have conflicts in their self-concepts that don't bother them at all," Lieberman comments. "Little kids don't try to tell themselves, 'I'm still the same person.' They just don't seem to connect up the little pieces of the self-concept."

Lieberman and his colleagues wondered if they could track children's changing self-concept with brain imaging. They have begun studying a group of children and plan to scan them every 18 months from ages nine to 15. "We asked kids to think about themselves and to think about Harry Potter," he says. He and his team have compared the brain activity in each task and compared the results with those in adults.

"When you look at 10-year-olds, they show this same medial prefrontal cortex activation as adults do," Lieberman notes. But another region that becomes active in adults, known as the precuneus, is a different story. "When they think about themselves, they activate this region *less* than they do when they think about Harry Potter." Lieberman suspects that in children, the self-network is still coming online. "They've got the stuff, but they're not applying it like adults do."

INSIGHTS INTO ALZHEIMER'S

Once the self-network does come online, however, it works very hard. "Even with the visual system, I can close my eyes and give it something of a rest," comments William Seeley, a neurologist at the University of California, San Francisco. "But I can never get away from living in my body or representing the fact that I'm the same person I was 10 seconds or 10 years ago. I can never escape that, so that network must be busy."

The more energy that a cell consumes, the greater its risk of damaging itself with toxic by-products. Seeley suspects that the hardworking neurons in the self-network are particularly vulnerable to this damage over the life span. Their vulnerability, he argues, may help neurologists make sense of some brain disorders that erode the self. "It is curious that we can't find certain pathological changes of Alzheimer's or other dementias in nonhuman species," Seeley says.

According to Seeley, the results of recent brain imaging studies of the self agree with findings by him and others on people with Alzheimer's and other dementias. People with Alzheimer's develop tangled proteins in their neurons. Some of the first regions to be damaged are the hippocampus and precuneus, which are among the areas involved in autobiographical memories. "They help you bring images of your past and future into mind and play with them," Seeley notes. "People with Alzheimer's are just less able to move smoothly back and forth through time."

As agonizing as it may be for family members to watch a loved one succumb to Alzheimer's, other kinds of dementia can have even more drastic effects on the self. In a condition known as frontotemporal dementia, swaths of the frontal and temporal lobes degenerate. In many cases, the medial prefrontal cortex is damaged. As the disease begins to ravage the self-network, people may undergo strange changes in personality.

One patient, described by Seeley and others in the journal *Neurology* in 2001, had collected jewelry and fine crystal for much of her life before abruptly starting to gather stuffed animals at age 62. A lifelong conservative, she began to berate people in stores who were buying conservative books and declared that "Republicans should be taken off the earth." Other patients have suddenly converted to new religions or become obsessed with painting or photography. Yet they have little insight into why they are no longer their old selves. "They say pretty shallow things, like 'This is just the way I am now,'" Seeley says. Within a few years, frontotemporal dementia can lead to death.

Michael Gazzaniga, director of Dartmouth's Center for Cognitive Neuroscience and a member of the President's Council on Bioethics, believes that deciphering the self may pose a new kind of ethical challenge. "I think there's going to be the working out of the circuits of self—self-referential memory, self-description, personality, self-awareness," Gazzaniga says. "There's going to be a sense of what has to be in place for the self to be active."

It is even possible, Gazzaniga suggests, that someday a brain scan might determine whether Alzheimer's or some other dementia has destroyed a person's self. "Someone's going to say, 'Where's Gramps?'" he predicts. "And they're going to be able to take a picture of Gramps under certain conditions and say, 'Those circuits are not working.'"

Gazzaniga wonders whether people will begin to consider the loss of the self when they write out their living wills. "Advance directives will come into play," he predicts. "The issue will be whether you deliver health care. If people catch pneumonia, do you give them antibiotics or let them go?"

Seeley offers a more conservative forecast, arguing that a brain scan on its own probably will not change people's minds about life-and-death decisions. Seeley thinks the real value of the science of the self will come in treatments for Alzheimer's and other dementias. "Once we know which brain regions are involved in self-representation, I think we can take an even closer look at which cells in that brain region are important and then look deeper and say which molecules within cells and which genes that govern them lead to this vulnerability," he says. "And if we've done that, we've gotten closer to disease mechanisms and cures. That's the best reason to study all this. It's not just to inform philosophers."

MORE TO EXPLORE

Feinberg, Todd E., and Julian Paul Keenan, eds. 2005. *The lost self: Pathologies of the brain and identity.* Oxford University Press.

Gillihan, Seth J., and Martha J. Farah. 2005. Is self special? A critical review of evidence from experimental psychology and cognitive neuroscience. *Psychological Bulletin* 131 (1): 76–97.

Lieberman, Matthew D., and Naomi I. Eisenberger. In press. Conflict and habit: A social cognitive neuroscience approach to the self. In *Psychological perspectives on self and identity,* vol. 4, ed. A. Tesser, J. V. Wood, and D. A. Stapel. American Psychological Association. Available online at www.scn.ucla.edu/pdf/rt4053_c004Lieberman.pdf.

Macrae, Neil, Todd F. Heatherton, and William M. Kelly. 2004. A self less ordinary: The medial prefrontal cortex and you. In *Cognitive neurosciences,* vol. 3, ed. Michael S. Gazzaniga. MIT Press.

CARL ZIMMER is a journalist based in Connecticut. His latest book, *Soul Made Flesh: The Discovery of the Brain—and How It Changed the World,* was recently published in paperback. He also writes The Loom, a blog about biology.
Originally published in *Scientific American,* Vol. 293, No. 5, November 2005.

5

How the Brain Creates the Mind

We have long wondered how the conscious mind comes to be. Greater understanding of brain function ought to lead to an eventual solution

ANTONIO R. DAMASIO

At the start of the new millennium, it is apparent that one question towers above all others in life sciences: How does the set of processes we call mind emerge from the activity of the organ we call brain? The question is hardly new. It has been formulated in one way or another for centuries. Once it became possible to pose the question and not be burned at the stake, it has been asked openly and insistently. Recently the question has preoccupied both the experts—neuroscientists, cognitive scientists and philosophers—and others who wonder about the origin of the mind, specifically the conscious mind.

The question of consciousness now occupies center stage because biology in general and neuroscience in particular have been so remarkably successful at unraveling a great many of life's secrets. More may have been learned about the brain and the mind in the 1990s—the so-called Decade of the Brain—than during the entire previous history of psychology and neuroscience. Elucidating the neurobiological basis of the conscious mind—a version of the classic mind-body problem—has become almost a residual challenge.

Contemplation of the mind may induce timidity in the contemplator, especially when consciousness becomes the focus of the inquiry. Some thinkers, expert and amateur alike, believe the question may be unanswerable in principle. For others, the relentless and exponential increase in new knowledge may give rise to a vertiginous feeling that no problem can resist the assault of science if only the theory is right and the techniques are powerful enough. The debate is intriguing and even unexpected, as no comparable doubts have been raised over the likelihood of explaining how the brain is responsible for processes such as vision or memory, which are obvious components of the larger process of the conscious mind.

I am firmly in the confident camp: a substantial explanation for the mind's emergence from the brain will be produced and perhaps soon. The giddy feeling, however, is tempered by the acknowledgment of some sobering difficulties.

Nothing is more familiar than the mind. Yet the pilgrim in search of the sources and mechanisms behind the mind embarks on a journey into a strange and exotic landscape. In no particular order, what follows are the main problems facing those who seek the biological basis for the conscious mind.

The first quandary involves the perspective one must adopt to study the conscious mind in relation to the brain in which we believe it originates. Anyone's body and brain are observable to third parties; the mind, though, is observable only to its owner. Multiple individuals confronted with the same body or brain can make the same observations of that body or brain, but no comparable direct third-person observation is possible for anyone's mind. The body and its brain are public, exposed, external and unequivocally objective entities. The mind is a private, hidden, internal, unequivocally subjective entity.

How and where then does the dependence of a first-person mind on a third-person body occur precisely? Techniques used to study the brain include refined brain scans and the measurement of patterns of activity in the brain's neurons. The naysayers argue that the exhaustive compilation of all these data adds up to *correlates* of mental states but nothing resembling an *actual mental state*. For them, detailed observation of living matter thus leads not to mind but simply to the details of living matter. The understanding of how living matter generates the sense of self that is the hallmark of a conscious mind—the sense that the images in my mind are mine and are formed in my perspective—is simply not possible. This argument, though incorrect, tends to silence most hopeful investigators of the conscious mind.

To the pessimists, the conscious-mind problem seems so intractable that it is not even possible to explain why the mind is even *about* something—why mental processes represent internal states or interactions with external objects. (Philosophers refer to this representational quality of the mind with the confusing term "intentionality.") This argument is false.

The final negative contention is the reminder that elucidating the emergence of the conscious mind depends on the existence of that same conscious mind. Conducting an investigation with the very instrument being investigated makes both the definition of the problem and the approach to a solution especially complicated. Given the conflict between observer and observed, we are told, the human intellect is unlikely to be up to the task of comprehending how mind emerges from brain. This conflict is real, but the notion that it is insurmountable is inaccurate.

In summary, the apparent uniqueness of the conscious-mind problem and the difficulties that complicate ways to get at that problem generate two effects: they frustrate those researchers committed to finding a solution and confirm the conviction of others who intuitively believe that a solution is beyond our reach.

EVALUATING THE DIFFICULTIES

Those who cite the inability of research on the living matter of the brain to reveal the "substance of mind" assume that the current knowledge of that living matter is sufficient to make such judgment final. This notion is entirely unacceptable. The current description of neurobiological phenomena is quite incomplete, any way you slice it. We have yet to resolve numerous details about the function of neurons and circuits at the molecular level; we do not yet grasp the behavior of populations of neurons within a local brain region; and our understanding of the large-scale systems made up of multiple brain regions is also incomplete. We are barely beginning to address the fact that interactions among many non-contiguous brain regions probably yield highly complex biological states that are vastly more than the sum of their parts.

In fact, the explanation of the physics related to biological events is still incomplete. Consequently, declaring the conscious-mind problem insoluble because we have studied the brain to the hilt and have not found the mind is ludicrous. We have not yet fully studied either neurobiology or its related physics. For example, at the finest level of description of mind, the swift construction, manipulation and superposition of many sensory images might require explanation at the quantum level. Incidentally, the notion of a possible role for quantum physics in the elucidation of mind, an idea usually associated with mathematical physicist Roger Penrose of the University of Oxford, is not an endorsement of his specific proposals, namely that consciousness is based on quantum-level phenomena occurring in the microtubules—constituents of neurons and other cells. The quantum level of operations might help explain how we have a mind, but I regard it as unnecessary to explain how we know that we own that mind—the issue I regard as most critical for a comprehensive account of consciousness.

The strangeness of the conscious-mind problem mostly reflects ignorance, which limits the imagination and has the curious effect of making the possi-

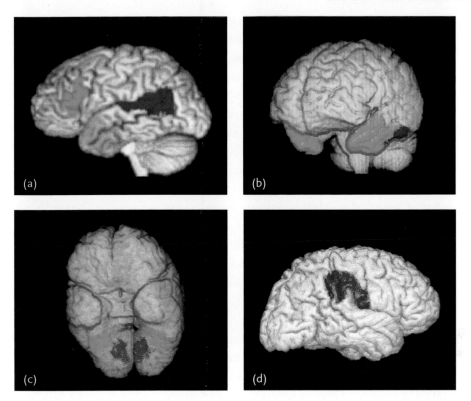

Neuroscience continues to associate specific brain structures with specific tasks. Some language regions are highlighted in images a and b. Color-processing (red) and face-processing (green) regions are shown in image c. One's own body sense depends on the region shown in image d.

ble seem impossible. Science-fiction writer Arthur C. Clarke has said, "Any sufficiently advanced technology is indistinguishable from magic." The "technology" of the brain is so complex as to appear magical, or at least unknowable. The appearance of a gulf between mental states and physical/biological phenomena comes from the large disparity between two bodies of knowledge—the good understanding of mind we have achieved through centuries of introspection and the efforts of cognitive science versus the incomplete neural specification we have achieved through the efforts of neuroscience. But there is no reason to expect that neurobiology cannot bridge the gulf. Nothing indicates that we have reached the edge of an abyss that would separate, in principle, the mental from the neural.

Therefore, I contend that the biological processes now presumed to correspond to mind processes in fact are mind processes and will be seen to be so when understood in sufficient detail. I am not denying the existence of the mind or saying that once we know what we need to know about biology the mind ceases to exist. I simply believe that the private, personal mind, precious and unique, indeed is biological and will one day be described in terms both biological and mental.

The other main objection to an understanding of mind is that the real conflict between observer and observed makes the human intellect unfit to study itself. It is important, however, to point out that the brain and mind are not a monolith: they have multiple structural levels, and the highest of those levels creates instruments that permit the observation of the other levels. For example, language endowed the mind with the power to categorize and manipulate knowledge according to logical principles, and that helps us classify observations as true or false. We should be modest about the likelihood of ever observing our entire nature. But declaring defeat before we even make the attempt defies Aristotle's observation that human beings are infinitely curious about their own nature.

REASONS FOR OPTIMISM

My proposal for a solution to the conundrum of the conscious mind requires breaking the problem into two parts. The first concern is how we generate what I call a "movie-in-the-brain." This "movie" is a metaphor for the integrated and unified composite of diverse sensory images—visual, auditory, tactile, olfactory and others—that constitutes the multimedia show we call mind. The second issue is the "self" and how we automatically generate a sense of ownership for the movie-in-the-brain. The two parts of the problem are related, with the latter nested in the former. Separating them is a useful research strategy, as each requires its own solution.

Neuroscientists have been attempting unwittingly to solve the movie-in-the-brain part of the conscious-mind problem for most of the history of the field. The endeavor of mapping the brain regions involved in constructing the movie began almost a century and a half ago, when Paul Broca and Carl Wernicke first suggested that different regions of the brain were involved in processing different aspects of language. More recently, thanks to the advent of ever more sophisticated tools, the effort has begun to reap handsome rewards.

Researchers can now directly record the activity of a single neuron or group of neurons and relate that activity to aspects of a specific mental state, such as the perception of the color red or of a curved line. Brain imaging techniques such as PET (positron-emission tomography) scans and fMR (functional magnetic resonance) scans reveal how different brain regions in a normal, living

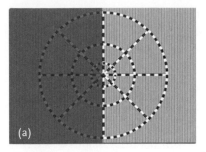

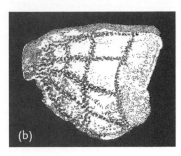

The brain's business is representing other things. Studies with macaques show a remarkable fidelity between a seen shape (a) and the shape of the neural activity pattern (b) in one of the layers of the primary visual cortex.

person are engaged by a certain mental effort, such as relating a word to an object or learning a particular face. Investigators can determine how molecules within microscopic neuron circuits participate in such diverse mental tasks, and they can identify the genes necessary for the production and deployment of those molecules.

Progress in this field has been swift ever since David H. Hubel and Torsten Wiesel of Harvard University provided the first clue for how brain circuits represent the shape of a given object, by demonstrating that neurons in the primary visual cortex were selectively tuned to respond to edges oriented in varied angles. Hubel and Margaret S. Livingstone, also at Harvard, later showed that other neurons in the primary visual cortex respond selectively to color but not shape. And Semir Zeki of University College London found that brain regions that received sensory information after the primary visual cortex did were specialized for the further processing of color or movement. These results provided a counterpart to observations made in living neurological patients: damage to distinct regions of the visual cortices interferes with color perception while leaving discernment of shape and movement intact.

A large body of work, in fact, now points to the existence of a correspondence between the structure of an object as taken in by the eye and the pattern of neuron activity generated within the visual cortex of the organism seeing that object.

Further remarkable progress involving aspects of the movie-in-the-brain has led to increased insights related to mechanisms of learning and memory. In rapid succession, research has revealed that the brain uses discrete systems for different types of learning. The basal ganglia and cerebellum are critical for the acquisition of skills—for example, learning to ride a bicycle or play a musical instrument. The hippocampus is integral to the learning of facts pertaining to such entities as people, places or events. And once facts are learned, the long-term memory of those facts relies on multicomponent brain systems, whose key parts are located in the vast brain expanses known as cerebral cortices.

Moreover, the process by which newly learned facts are consolidated in long-term memory goes beyond properly working hippocampi and cerebral cortices. Certain processes must take place, at the level of neurons and molecules, so that the neural circuits are etched, so to speak, with the impressions of a newly learned fact. This etching depends on strengthening or weakening the contacts between neurons, known as synapses. A provocative finding by Eric R. Kandel of Columbia University and Timothy P. Tully of Cold Spring Harbor Laboratory is that etching the impression requires the synthesis of fresh proteins, which in turn relies on the engagement of specific genes within the neurons charged with supporting the consolidated memory.

These brief illustrations of progress could be expanded with other revelations from the study of language, emotion and decision making. Whatever mental function we consider, it is possible to identify distinct parts of the brain that contribute to the production of a function by working in concert; a close correspondence exists between the appearance of a mental state or behavior and the activity of selected brain regions. And that correspondence can be established between a given macroscopically identifiable region (for example, the primary visual cortex, a language-related area or an emotion-related nucleus) and the microscopic neuron circuits that constitute the region.

Most exciting is that these impressive advances in the study of the brain are a mere beginning. New analytical techniques continuously improve the ability to study neural function at the molecular level and to investigate the highly complex large-scale phenomena arising from the whole brain. Revelations from those two areas will make possible ever finer correspondences between brain states and mental states, between brain and mind. As technology develops and the ingenuity of researchers grows, the fine grain of physical structures and biological activities that constitute the movie-in-the-brain will gradually come into focus.

CONFRONTING THE SELF

The momentum of current research on cognitive neuroscience, and the sheer accumulation of powerful facts, may well convince many doubters that the neural basis for the movie-in-the-brain can be identified. But the skeptics will still find it difficult to accept that the second part of the conscious-mind problem—the emergence of a sense of self—can be solved at all. Although I grant that solving this part of the problem is by no means obvious, a possible solution has been proposed, and a hypothesis is being tested.

The main ideas behind the hypothesis involve the unique representational ability of the brain. Cells in the kidney or liver perform their assigned functional roles and do not represent any other cells or functions. But brain cells, at every level of the nervous system, represent entities or events occurring elsewhere in the organism. Brain cells are assigned by design to be *about* other things and other doings. They are born cartographers of the geography of an organism and of the events that take place within that geography. The oft-quoted mystery of the "intentional" mind relative to the representation of external objects turns out to be no mystery at all. The philosophical despair that surrounds this "intentionality" hurdle alluded to earlier—why mental states represent internal emotions or interactions with external objects—lifts with the consideration of the brain in a Darwinian context: evolution has crafted a brain that is in the business of directly representing the organism and indirectly representing whatever the organism interacts with.

The brain's natural intentionality then takes us to another established fact: the brain possesses devices within its structure that are designed to manage the life of the organism in such a way that the internal chemical balances indispensable for survival are maintained at all times. These devices are neither hypothetical nor abstract; they are located in the brain's core, the brain stem and hypothalamus. The brain devices that regulate life also represent, of necessity, the constantly changing states of the organism as they occur. In other words, the brain has a natural means to represent the structure and state of the *whole* living organism.

But how is it possible to move from such a biological self to the sense of ownership of one's thoughts, the sense that one's thoughts are constructed in one's own perspective, without falling into the trap of invoking an all-knowing homunculus who interprets one's reality? How is it possible to know about self and surroundings? I have argued in my book *The Feeling of What Happens* that the biological foundation for the sense of self can be found in those brain devices that represent, moment by moment, the continuity of the same individual organism.

Simply put, my hypothesis suggests that the brain uses structures designed to map both the organism and external objects to create a fresh, second-order representation. This representation indicates that the organism, as mapped in the brain, is involved in interacting with an object, also mapped in the brain. The second-order representation is no abstraction; it occurs in neural structures such as the thalamus and the cingulate cortices.

Such newly minted knowledge adds important information to the evolving mental process. Specifically, it *presents* within the mental process the information that the organism is the owner of the mental process. It volunteers an answer to a question never posed: To whom is this happening? The sense of a self in the act of knowing is thus created, and that forms the basis for the first-person perspective that characterizes the conscious mind.

Again from an evolutionary perspective, the imperative for a sense of self becomes clear. As Willy Loman's wife says in Arthur Miller's *Death of a Salesman:* "Attention must be paid!" Imagine a self-aware organism versus the same type of organism lacking it. A self-aware organism has an incentive to heed the alarm signals provided by the movie-in-the-brain (for instance, pain caused by a particular object) and plan the future avoidance of such an object. Evolution of self rewards awareness, which is clearly a survival advantage.

With the movie metaphor in mind, if you will, my solution to the conscious-mind problem is that the sense of self in the act of knowing emerges *within* the movie. Self-awareness is actually part of the movie and thus creates, within the same frame, the "seen" and the "seer," the "thought" and the "thinker." There is no separate spectator for the movie-in-the-brain. The idea of spectator is constructed within the movie, and no ghostly homunculus haunts the theater. Objective brain processes knit the subjectivity of the conscious mind out of the cloth of sensory mapping. And because the most fundamental sensory mapping pertains to body states and is imaged as feelings, the sense of self in the act of knowing emerges as a special kind of feeling—the feeling of what happens in an organism caught in the act of interacting with an object.

THE FUTURE

I would be foolish to make predictions about what can and cannot be discovered or about when something might be discovered and the route of a discovery. Nevertheless, it is probably safe to say that by 2050 sufficient knowledge of biological phenomena will have wiped out the traditional dualistic separations of body/brain, body/mind and brain/mind.

Some observers may fear that pinning down the physical structure of something as precious and dignified as the human mind may result in its being downgraded or vanishing entirely. But explaining the origins and workings of the mind in biological tissue will not do away with the mind, and the awe we have for it can be extended to the amazing microstructure of the organism and to the immensely complex functions that allow such a microstructure to generate the mind. By understanding the mind at a deeper level, we will see it as nature's most complex set of biological phenomena rather than as a mystery with an unknown nature. The mind will survive explanation, just as a rose's perfume, its molecular structure deduced, will still smell as sweet.

MORE TO EXPLORE

Churchland, Paul M. 1995. *The engine of reason, the seat of the soul: A philosophical journey into the brain*. MIT Press.

Damasio, Antonio R. 1999. The feeling of what happens: Body and emotion in the making of consciousness. Harcourt Brace.

————. 2003. *Looking for Spinoza: Joy, sorrow and the human brain*. Harcourt.

Dennett, Daniel C. 1996. *Consciousness explained*. Little, Brown.

Hubel, David H. 1995. *Eye, brain, and vision*. Scientific American Library, no. 22. W. H. Freeman.

ANTONIO R. DAMASIO is M. W. Van Allen Distinguished Professor and head of the Department of Neurology at the University of Iowa College of Medicine and adjunct professor at the Salk Institute for Biological Studies in San Diego. He was born in Portugal and received his M.D. and Ph.D. from the University of Lisbon. With his wife, Hanna, Damasio created a facility at Iowa dedicated to the investigation of neurological disorders of mind and behavior. A member of the Institute of Medicine of the National Academy of Sciences and of the American Academy of Arts and Sciences, Damasio is the author of *Descartes' Error: Emotion, Reason, and the Human Brain* (1994), *The Feeling of What Happens: Body and Emotion in the Making of Consciousness* (1999) and *Looking for Spinoza* (2003).

Originally published in *The Hidden Mind*, special edition of *Scientific American*, August 2002.

6

The New Science of Mind

A forecast of the major problems scientists need to solve

ERIC R. KANDEL

Understanding the human mind in biological terms has emerged as the central challenge for science in the 21st century. We want to understand the biological nature of perception, learning, memory, thought, consciousness and the limits of free will. That biologists would be in a position to explore these mental processes was unthinkable even a few decades ago. Until the middle of the 20th century, when I began my career as a neuroscientist, the idea that mind, the most complex set of processes in the universe, might yield its deepest secrets to biological analysis and perhaps do this on the molecular level could not be entertained seriously.

The dramatic achievements of biology during the past 50 years have now made this possible. The discovery of the structure of DNA by James Watson and Francis Crick in 1953 revolutionized biology, giving it an intellectual framework for understanding how information from the genes controls the functioning of the cell. That discovery led to a basic understanding of how genes are regulated, how they give rise to the proteins that determine the functioning of cells, and how development turns genes and proteins on and off to establish the body plan of an organism. With these extraordinary accomplishments behind it, biology assumed a central position in the constellation of sciences, in parallel with physics and chemistry.

Imbued with new knowledge and confidence, biology turned its attention to its loftiest goal: understanding the biological nature of the human mind. This effort, long considered to be prescientific, is already in full swing. Indeed, when intellectual historians look back on the last two decades of the 20th century, they are likely to comment on the surprising fact that the most valuable insights into the human mind to emerge during this period did not come from the disciplines traditionally concerned with mind—philosophy, psychology or psychoanalysis. Instead they came from a merger of these disciplines with the biology of the brain, a new synthesis energized recently by dramatic achievements in molecular biology.

MIND IS BRAIN

The result has been a new science of mind, a science that uses the power of molecular biology to examine the great remaining mysteries of life. This new science is based on five principles. First, mind and brain are inseparable. The brain is a complex biological organ of great computational capability that constructs our sensory experiences, regulates our thoughts and emotions, and controls our actions. The brain is responsible not only for relatively simple motor behaviors, such as running and eating, but also for the complex acts that we consider quintessentially human, such as thinking, speaking and creating works of art. Looked at from this perspective, mind is a set of operations carried out by the brain, much as walking is a set of operations carried out by the legs, except dramatically more complex.

Second, each mental function in the brain—from the simplest reflex to the most creative acts in language, music and art—is carried out by specialized neural circuits in different regions of the brain. This is why it is preferable to use the term "biology of mind" to refer to the set of mental operations carried out by these specialized neural circuits rather than "biology of *the* mind," which connotes a place and implies a single brain location that carries out all mental operations.

Third, all of these circuits are made up of the same elementary signaling units, the nerve cells. Fourth, the neural circuits use specific molecules to generate signals within and between nerve cells. Finally, these specific signaling molecules have been conserved—retained, as it were—through millions of years of evolution. Some of them were present in the cells of our most ancient ancestors and can be found today in our most distant and primitive evolutionary relatives: single-celled organisms such as bacteria and yeast and simple multicellular organisms such as worms, flies and snails. These creatures use the same molecules to organize maneuvering through their environment that we use to govern our daily lives and adjust to our environment.

Thus, we gain from the new science of mind not only insights into ourselves—how we perceive, learn, remember, feel and act—but also a new perspective of ourselves in the context of biological evolution. It makes us appreciate that the human mind evolved from molecules used by our ancestors and that the extraordinary conservation of the molecular mechanisms that regulate life's various processes also applies to our mental life.

Because of its broad implications for individual and social well-being, there is now a general consensus in the scientific community that the biology of mind will be to the 21st century what the biology of the gene was to the 20th century.

A SYSTEMS APPROACH

As we enter the 21st century, the new science of mind faces remarkable challenges. Researchers of memory storage, including my colleagues and me, are only standing at the foothills of a great mountain range. We have learned about the cellular and molecular mechanisms of memory storage, but now we must progress to the systems properties of memory. For example, which neural circuits are critical for which kinds of memory?

How does the brain encode internal representations of a face, scene, melody or experience?

To move from where we are to where we want to be, we must undertake major conceptual shifts in how we think about the brain. One such shift involves moving away from elementary processes (that is, single proteins, genes and cells) and toward systems properties, such as complex networks of proteins or nerve cells, the functioning of whole organisms and the interaction of groups of organisms. Cellular and molecular approaches will no doubt continue to yield important information, but alone they cannot reveal the intricacies of internal representations within, or interactions among, neural circuits—the key steps linking cellular and molecular neuroscience to cognitive neuroscience.

To develop an approach that relates neural systems to complex cognitive functions, we must focus on neural circuits, discerning how patterns of activity in different circuits merge to form a coherent representation. To learn how we perceive and recall complex experiences, we must determine how neural networks are organized and how attention and awareness shape and reconfigure neural activity in those networks. To accomplish these goals, biology must focus more on human beings and on nonhuman primates using imaging techniques that can resolve the activity of individual neurons and neuronal networks.

WHAT IS ATTENTION?

These reflections have led me to wonder what scientific questions I would pursue were I to start anew. I have two requirements for selecting such a research problem. First, it must allow me to participate in opening a new area of research that will occupy me for a long time. (I like long-term commitments, not brief romances.) Second, the problem must lie at the intersection of two or more disciplines. Based on these criteria, three sets of questions appeal to me.

First, how does the brain process sensory information consciously, and how does conscious attention stabilize memory? Crick and California Institute of Technology neuroscientist Christof Koch have argued persuasively that selective attention is not only an important area of investigation in its own right but also a critical component of consciousness. Regarding attention, I would focus on "place cells," which determine an animal's location in space, in the hippocampus—a brain region linked with long-term memory—and how place cells create an enduring spatial map only when an organism focuses its spotlight of attention on its surroundings.

What is this spotlight of attention? How does it trigger the neural circuitry of spatial memory to encode information? Moreover, what modulatory brain systems turn on when an animal pays attention, and how are they activated? How does attention enable me to embark on "mental time travel" to the little apartment in Vienna where I grew up? To investigate these matters, we ought to extend our studies of memory beyond laboratory animals to human beings.

The second question is, How do unconscious and conscious mental processes relate to one another in people? The notion that we are unaware of much of our mental life—an idea that German physician and physicist Hermann von Helmholtz proposed in 1860—lies at the core of psychoanalysis. Only through understanding such issues can we address, in biologically meaningful terms, Sigmund Freud's theories proposed in 1899 about conscious and unconscious conflicts and memory. To Helmholtz's notion, Freud added the important observation that by paying attention, we can access some of our unconscious mental processes—ones that would otherwise go unnoticed.

UNCONSCIOUS MECHANISMS

Seen from this perspective—a view that most neural scientists now hold—most of our mental life is unconscious. And we become aware of many otherwise inaccessible brain processes only through words and images. So, in principle, we should be able to use brain-imaging techniques to connect psychoanalytic processes with brain anatomy and neural functioning. Such a bridge might enable us to learn how disease states alter unconscious processes and how psychotherapy might help reconfigure them. Unconscious psychic processes play such a large role in our lives; perhaps biology can help us learn about them.

The final question is, How can we link molecular biology of mind to sociology and thus develop a realistic molecular sociobiology? Several researchers have made a fine start toward this goal. For example, Cori Bargmann, a geneticist now at the Rockefeller University, has studied two variants of the soil nematode *Caenorhabditis elegans* that differ in their feeding patterns. One is solitary and seeks food alone. The other is social and forages in groups. The only difference between the two is one amino acid in an otherwise identical receptor protein. Transferring the receptor from a social worm to a solitary one causes the solitary creature to socialize.

Another example involves male courtship in the fruit fly *Drosophila*. A key protein, called Fruitless, governs this instinctive behavior, and Fruitless is expressed differently in male and female flies. Ebru Demir and Barry J. Dickson, neuroscientists at the Research Institute of Molecular Pathology in Vienna, have made the remarkable discovery that when female flies express the male form of this protein, they mount and direct courtship toward other female flies—or toward males genetically engineered to produce a characteristic female odor, or pheromone. Dickson also found that for *Drosophila* to grow the neural circuitry for courtship and sexual preference, the Fruitless gene must be present and active during the fly's early development. (If scientists add this gene later, instead, then it does not have the same effect.)

Still a third example comes from Giacomo Rizzolatti, a neuroscientist at the University of Parma in Italy. He discovered that certain neurons in the premotor cortex become active when a monkey carries out a specific action with its hand, such as putting a peanut in its mouth. Remarkably, the same neurons respond when a monkey watches another monkey (or even a person) put food in its mouth. Rizzolatti calls these cells "mirror neurons," suggesting that they offer insight into imitation, identification, empathy and possibly the ability to mime vocalization—all unconscious mental processes intrinsic to human interaction. Vilayanur S. Ramachandran, a neuroscientist at the University of California, San Diego, has found evidence of comparable neurons in the premotor cortex of humans.

MENTAL STATES ARE BRAIN STATES

Reflecting on just these three research strands, I can see whole new areas of biology opening up, providing a sense of what makes us social, communicative beings. An ambitious undertaking of this kind might not only reveal what enables members of a cohesive group to recognize one another but also give us insight into tribalism, which so often promotes fear, hatred and intolerance of outsiders.

Since the 1980s the path toward merging mind and brain research has become clearer. As a result, psychiatry has taken on a new role, both stimulating and benefiting from biological thought. During the past few years, even members of the psychoanalytic community have taken on a keen interest in the biology of mind, acknowledging that every mental state is a brain state, that all mental disorders involve disorders of brain function. Treatments work when they alter the brain's structure and functioning.

To give a sense of how attitudes among researchers have changed in recent years, when in 1962 I turned from studying the hippocampus in the mammalian brain to studying simple forms of learning in the sea slug *Aplysia*, I encountered many negative reactions. At the time, there was a strong sense among brain scientists that mammalian brains differed radically from those of lower vertebrates, such as fish and frogs—and were incomparably more complex than invertebrate brains. The fact that Nobel Prize-winning neuroscientists, such as the late Alan Hodgkin, Andrew F. Huxley of the University of Cambridge and the late Bernard Katz, discovered the fundamental principles of signaling in the human nervous system by probing the neural axons of squids and the synapses joining nerves and muscles in frogs seemed to these mammalian chauvinists an exception. Of course all nerve cells are similar, they conceded, but neural circuitry and behavior differ markedly in vertebrates and invertebrates.

OF FLIES AND MEN

This schism persisted until molecular biologists revealed the amazing continuity—throughout evolution, from lower to higher organisms—of the genes and proteins governing all neural systems. Even then, disputes arose as to whether the cellular and molecular mechanisms of learning and memory revealed in simple animal studies would generalize to more complex animals. Neuroscientists argued about whether elementary forms of learning such as sensitization and habituation gave rise to forms of memory that would be useful to study. The ethologists, who study animal behavior in natural settings, emphasized the generality of these simple forms of memory, whereas the behaviorists highlighted associative forms of learning, such as classical and operant conditioning, which are clearly more complex. The disagreements were eventually resolved in two ways. First, Seymour Benzer, a biologist at Caltech, proved that cyclic AMP (adenosine monophosphate), which is important for short-term sensitization in *Aplysia*, plays a critical role in more complex forms of learning, such as classical conditioning, in a more complex animal—namely, *Drosophila*. Second, the regulatory protein CREB, first identified in my laboratory in *Aplysia*, proved to be a key molecular component in switching from short- to long-term memory in many forms of learning and types of organisms, ranging from snails to flies, to mice and to people. Evidently, learning and memory, as well as synaptic and neuronal plasticity, or ability to change, involve a family of processes that vary in molecular subtleties but share various components and a common logic. In most cases, these discussions proved beneficial for science, sharpening questions and moving research forward. To me, the most important facet of the debates was the sense that we were progressing in the right direction.

These debates influenced my views, as did my psychiatric training and psychoanalytic interests, which lie at the very core of my scientific thinking. Together they shaped my perspective on mind and behavior, establishing overarching ideas that influenced nearly every aspect of my research and fueled my interest in conscious and unconscious memory.

PASSION AND BOLD DISCOVERIES

Few experiences excite and stimulate the imagination more than discovering something new, no matter how modest. A new finding allows someone to see for the first time a small piece of nature's puzzle. Becoming absorbed in a problem, I find it helpful to develop a comprehensive perspective by learning what previous scientists thought about it. I want to know not only which lines of thought proved productive but also where and why other lines proved unproductive. And so Freud, as well as other early researchers in learning and memory—such as the classic psychologists William James, Edward Thorndike, Ivan Pavlov, B. F. Skinner and Ulric Neisser of Cornell University—all strongly influenced my thought. Their thinking, and even the unproductive paths they followed, provided a rich cultural background for my later work. Thus, my initial aspirations in psychoanalysis were hardly a detour. Rather they became the educational bedrock of all that I have tried to learn.

It is important to be bold. One should tackle difficult problems, especially those that initially appear messy and unstructured. One should not fear trying new things, such as moving from one field to another or working at the boundaries of disciplines—where the most interesting problems often emerge. Most good scientists never hesitate to ask questions, explore unfamiliar terrain, follow their instincts or learn new science along the way. Nothing stimulates self-education more than pursuing a new area of research.

Defining a problem, or a set of interrelated problems, with a long trajectory is also critical for success. Early on I stumbled fortunately onto an interesting problem while studying the hippocampus and memory, then switched decisively to investigate learning in a simple animal. Both problems had enough intellectual sweep and scope to carry me through many experimental failures and disappointments.

As a result, I did not share the midcareer malaise of some colleagues who grew bored with their science and turned to other things. Rather I thrived on testing new ideas. My friend and colleague Richard Axel, a fellow neuroscientist at Columbia University who received the Nobel Prize in 2004 for his remarkable discovery that there are 1,000 different receptors for smell, often speaks about the addictive quality of reviewing in one's mind new and interesting findings. Unless Richard sees new data coming along, he becomes despondent—a feeling many of us share. Such is the constructive side of addiction, one that I have experienced in pursuit of my own lifelong passion—namely, to understand the cellular and molecular basis of mind.

MORE TO EXPLORE

Kandel, Eric R. 2005. *Psychiatry, psychoanalysis, and the new biology of mind.* American Psychiatric Publishing.

———. 2006. *In search of memory: The emergence of a new science of mind.* W. W. Norton.

Kandel, Eric R., James H. Schwartz, and Thomas M. Jessell. 2000. *Principles of neural science.* McGraw-Hill Medical.

Squire, Larry R., and Eric R. Kandel. 1999. *Memory: From mind to molecules.* Scientific American Books.

ERIC R. KANDEL is University Professor at Columbia University; Fred Kavli Professor and director, Kavli Institute for Brain Sciences; and a senior investigator at the Howard Hughes Medical Institute. In 2000 he received the Nobel Prize in Physiology or Medicine. This article is adapted with permission from Kandel's new book, *In Search of Memory: The Emergence of a New Science of Mind* (W. W. Norton & Company, 2006). Originally published in *Scientific American Mind*, Vol. 17, No. 2, April/May 2006.

Part 2 Matter

7

Vision: A Window on Consciousness

In their search for the mind, scientists are focusing on visual perception—how we interpret what we see

NIKOS K. LOGOTHETIS

When you first look at the center image in the painting by Salvador Dalí reproduced on page 80, what do you see? Most people immediately perceive a man's face, eyes gazing skyward and lips pursed under a bushy mustache. But when you look again, the image rearranges itself into a more complex tableau. The man's nose and white mustache become the mobcap and cape of a seated woman. The glimmers in the man's eyes reveal themselves as lights in the windows— or glints on the roofs—of two cottages nestled in darkened hillsides. Shadows on the man's cheek emerge as a child in short pants standing beside the seated woman—both of whom, it is now clear, are looking across a lake at the cottages from a hole in a brick wall, a hole that we once saw as the outline of the man's face.

In 1940, when he rendered *Old Age, Adolescence, Infancy (The Three Ages)*— which contains three "faces"—Dalí was toying with the capacity of the viewer's mind to interpret two different images from the same set of brushstrokes. More than 50 years later, researchers, including my colleagues and me, are using similarly ambiguous visual stimuli to try to identify the brain activity that underlies consciousness. Specifically, we want to know what happens in the brain at the instant when, for example, an observer comprehends that the three faces in Dalí's picture are not really faces at all.

Consciousness is a difficult concept to define, much less to study. Neuroscientists have in recent years made impressive progress toward understanding the complex patterns of activity that occur in nerve cells, or neurons, in the brain. Even so, most people, including many scientists, still find the notion that electrochemical discharges in neurons can explain the mind—and in particular consciousness—challenging.

Yet, as Nobel laureate Francis Crick of the Salk Institute for Biological Studies in San Diego and Christof Koch of the California Institute of Technology have argued, the problem of consciousness can be broken down into several separate questions, some of which can be subjected to scientific inquiry. For example, rather than worrying about what consciousness is, one can ask: What is the difference between the neural processes that correlate with a particular conscious experience and those that do not?

NOW YOU SEE IT . . .

That is where ambiguous stimuli come in. Perceptual ambiguity is not a whimsical behavior specific to the organization of the visual system. Rather it tells us something about the organization of the entire brain and its way of making us aware of all sensory information. Take, for instance, the meaningless string of French words *pas de lieu Rhône que nous*, cited by the psychologist William James in 1890. You can read this over and over again without recognizing that it sounds just like the phrase "paddle your own canoe." What changes in neural activity occur when the meaningful sentence suddenly reaches consciousness?

In our work with ambiguous visual stimuli, we use images that not only give rise to two distinct perceptions but also instigate a continuous alternation be-

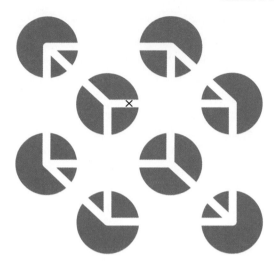

Necker cube can be viewed two different ways, depending on whether you see the "x" on the top front edge of the cube or on its rear face. Sometimes the cube appears superimposed on the circles; other times it seems as if the circles are holes and the cube is floating behind the page.

tween the two. A familiar example is the Necker cube. This figure is perceived as a three-dimensional cube, but the apparent perspective of the cube appears to shift every few seconds. Obviously, this alternation must correspond to something happening in the brain.

A skeptic might argue that we sometimes perceive a stimulus without being truly conscious of it, as when, for example, we "automatically" stop at a red light when driving. But the stimuli and the situations that I investigate are actually designed to reach consciousness.

We know that our stimuli reach awareness in human beings, because they can tell us about their experience. But it is not usually possible to study the activity of individual neurons in awake humans, so we perform our experiments with alert monkeys that have been trained to report what they are perceiving by pressing levers or by looking in a particular direction. Monkeys' brains are organized like those of humans, and they respond to such stimuli much as humans do. Consequently, we think the animals are conscious in somewhat the same way as humans are.

We investigate ambiguities that result when two different visual patterns are presented simultaneously to each eye, a phenomenon called binocular rivalry.

HOW TO EXPERIENCE BINOCULAR RIVALRY

To simulate binocular rivalry at home, use your right hand to hold the cardboard cylinder from a roll of paper towels (or a piece of paper rolled into a tube) against your right eye. Hold your left hand, palm facing you, roughly four inches in front of your left eye, with the edge of your hand touching the tube.

At first it will appear as though your hand has a hole in it, as your brain concentrates on the stimulus from your right eye. After a few seconds, though, the "hole" will fill in with a fuzzy perception of your whole palm from your left eye. If you keep looking, the two images will alternate, as your brain selects first the visual stimulus viewed by one eye, then that viewed by the other. The alternation is, however, a bit biased; you will probably perceive the visual stimulus you see through the cylinder more frequently than you will see your palm.

The bias occurs for two reasons. First, your palm is out of focus because it is much closer to your face, and blurred visual stimuli tend to be weaker competitors in binocular rivalry than sharp patterns, such as the surroundings you are viewing through the tube. Second, your palm is a relatively smooth surface with less contrast and fewer contours than your comparatively rich environment. In the laboratory, we carefully select the patterns viewed by the subjects to eliminate such bias. —N.K.L.

When people are put in this situation, their brains become aware first of one perception and then the other, in a slowly alternating sequence.

In the laboratory, we use stereoscopes to create this effect. Trained monkeys exposed to such visual stimulation report that they, too, experience a perception that changes every few seconds. Our experiments have enabled us to trace neural activity that corresponds to these changing reports.

IN THE MIND'S EYE

Studies of neural activity in animals conducted over several decades have established that visual information leaving the eyes ascends through successive stages of a neural data-processing system. Different modules analyze various attributes of the visual field. In general, the type of processing becomes more specialized the farther the information moves along the visual pathway.

At the start of the pathway, images from the retina at the back of each eye are channeled first to a pair of small structures deep in the brain called the lateral geniculate nuclei (LGN). Individual neurons in the LGN can be activated by visual stimulation from either one eye or the other but not both. They respond to any change of brightness or color in a specific region within an area of view known as the receptive field, which varies among neurons.

From the LGN, visual information moves to the primary visual cortex, known as VI, which is at the back of the head. Neurons in VI behave differently than those in the LGN do. They can usually be activated by either eye, but they are also sensitive to specific attributes, such as the direction of motion of a stimulus placed within their receptive field. Visual information is transmitted from VI to more than two dozen other distinct cortical regions.

Some information from VI can be traced as it moves through areas known as V2 and V4 before winding up in regions known as the inferior temporal cortex (ITC), which like all the other structures are bilateral. A large number of investigations, including neurological studies of people who have experienced brain damage, suggest that the ITC is important in perceiving form and recognizing objects. Neurons in V4 are known to respond selectively to aspects of visual stimuli critical to discerning shapes. In the ITC, some neurons behave like V4 cells, but others respond only when entire objects, such as faces, are placed within their very large receptive fields.

Other signals from VI pass through regions V2, V3 and an area known as MT/V5 before eventually reaching a part of the brain called the parietal lobe. Most neurons in MT/V5 respond strongly to items moving in a specific direction. Neurons in other areas of the parietal lobe respond when an animal pays attention to a stimulus or intends to move toward it.

One surprising observation made in early experiments is that many neurons in these visual pathways, both in VI and in higher levels of the processing hierarchy, still respond with their characteristic selectivity to visual stimuli even in animals that have been completely anesthetized. Clearly, an animal (or a human) is not conscious of all neural activity.

The observation raises the question of whether awareness is the result of the activation of special brain regions or clusters of neurons. The study of binocular rivalry in alert, trained monkeys allows us to approach that question, at least to some extent. In such experiments, a researcher presents each animal with a variety of visual stimuli, usually patterns or figures projected onto a screen. Monkeys can easily be trained to report accurately what stimulus they perceive by means of rewards of fruit juice.

During the experiment, the scientist uses electrodes to record the activity of neurons in the visual-processing pathway. Neurons vary markedly in their responsiveness when identical stimuli are presented to both eyes simultaneously. Stimulus pattern A might provoke activity in one neuron, for instance, whereas stimulus pattern B does not.

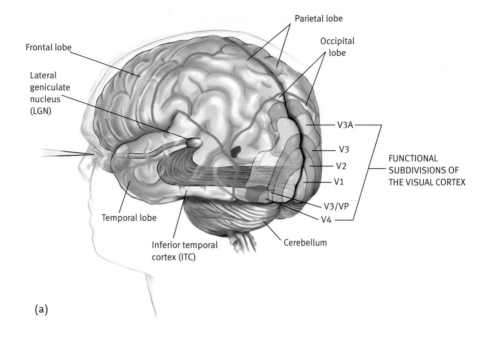

Frontal lobe

Lateral
geniculate
nucleus
(LGN)

Parietal lobe

Occipital
lobe

V3A
V3
V2
V1

FUNCTIONAL
SUBDIVISIONS OF
THE VISUAL CORTEX

V3/VP
V4

Temporal lobe

Inferior temporal
cortex (ITC)

Cerebellum

(a)

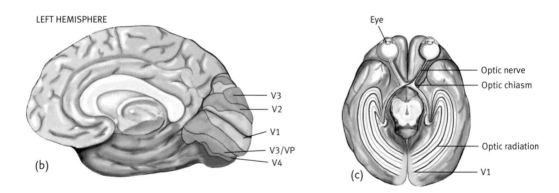

LEFT HEMISPHERE

V3
V2

V1

V3/VP
V4

(b)

Eye

Optic nerve
Optic chiasm

Optic radiation

V1

(c)

Human visual pathway begins with the eyes and extends through several interior brain structures before ascending to the various regions of the visual cortex (V1, and so on). At the optic chiasm, the optic nerves cross over partially so that each hemisphere of the brain receives input from both eyes. The information is filtered by the lateral geniculate nucleus, which consists of layers of nerve cells that each respond only to stimuli from one eye. The inferior temporal cortex is important for seeing forms. Some cells from each area are active only when a person or monkey becomes conscious of a given stimulus.

Once an experimenter has identified an effective and an ineffective stimulus for a given neuron (by presenting the same stimulus to both eyes at once), the two stimuli can be presented so that a different one is seen by each eye. We expect that, like a human in this situation, the monkey will become aware of the two stimuli in an alternating sequence. And, indeed, that is what the monkeys tell us by their responses when we present them with such rivalrous pairs of stimuli. By recording from neurons during successive presentations of rivalrous pairs, an experimenter can evaluate which neurons change their activity only when the stimuli change and which neurons alter their rate of firing when the animal reports a changed perception that is not accompanied by a change in the stimuli.

Jeffrey D. Schall, now at Vanderbilt University, and I carried out a version of this experiment in which one eye saw a grating that drifted slowly upward while the other eye saw a downward-moving grating. We recorded from visual area MT/V5, where cells tend to be responsive to motion. We found that about 43 percent of the cells in this area changed their level of activity when the monkey indicated that its perception had changed from up to down, or vice versa. Most of these cells were in the deepest layers of MT/V5.

The percentage we measured was actually a lower proportion than most scientists would have guessed, because almost all neurons in MT/V5 are sensitive to direction of movement. The majority of neurons in MT/V5 did behave somewhat like those in V1, remaining active when their preferred stimulus was in view of either eye, whether it was being perceived or not.

There were further surprises. Some 11 percent of the neurons examined were excited when the monkey reported perceiving the more effective stimulus of an upward/downward pair for the neuron in question. But, paradoxically, a similar proportion of neurons was most excited when the most effective stimulus was not perceived—even though it was in clear view of one eye. Other neurons could not be categorized as preferring one stimulus over another.

While we were both at Baylor College of Medicine, David A. Leopold and I studied neurons in parts of the brain known to be important in recognizing objects. (Leopold is now with me at the Max Planck Institute for Biological Cybernetics in Tübingen, Germany.) We recorded activity in V4, as well as in V1 and V2, while animals viewed stimuli consisting of lines sloping either to the left or to the right. In V4 the proportion of cells whose activity reflected perception was similar to that which Schall and I had found in MT/V5, about 40 percent. But again, a substantial proportion fired best when their preferred stimulus was not perceived. In V1 and V2, in contrast, fewer than one in 10 of the cells fired exclusively when their more effective stimulus was perceived, and none did so when it was not perceived.

The pattern of activity was entirely different in the ITC. David L. Sheinberg, now at Brown University, and I recorded from this area after training monkeys to report their perceptions during rivalry between complex visual patterns, such as images of humans, animals and various man-made objects. We found that almost all neurons, about 90 percent, responded vigorously when their preferred pattern was perceived but that their activity was profoundly inhibited when this pattern was not being experienced.

So it seems that by the time visual signals reach the ITC, the great majority of neurons are responding in a way that is linked to perception. Frank Tong, Ken Nakayama and Nancy Kanwisher of Harvard University have used functional magnetic resonance imaging (fMRI)—which yields pictures of brain activity by measuring increases in blood flow in specific areas of the brain—to study people experiencing binocular rivalry. They found that the ITC was particularly active when the subjects reported that they were seeing images of faces.

In short, most of the neurons in the earlier stages of the visual pathway responded mainly to whether their preferred visual stimulus was in view or not, although a few showed behavior that could be related to changes in the animal's perception. In the later stages of processing, on the other hand, the propor-

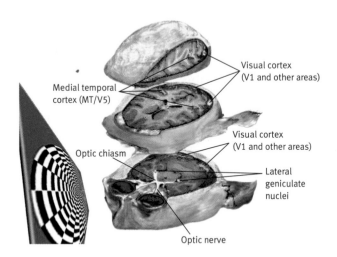

Images of brain activity are from an anesthetized monkey that was presented with a rotating, high-contrast visual stimulus (lower left). These views, taken using functional magnetic resonance imaging, show that even though the monkey is unconscious, its vision-processing areas—including the lateral geniculate nuclei (LGN), primary visual cortex (V1) and medial temporal cortex (MT/V5)—are busy.

tion whose activity reflected the animal's perception increased until it reached 90 percent.

A critic might object that the changing perceptions that monkeys report during binocular rivalry could be caused by the brain suppressing visual information at the start of the visual pathway, first from one eye and then from the other, so that the brain perceives a single image at any given time. If that were happening, changing neural activity and perceptions would simply represent the result of input that had switched from one eye to the other and would not be relevant to visual consciousness in other situations. But experimental evidence shows decisively that input from both eyes is continuously processed in the visual system during binocular rivalry.

We know this because it turns out that in humans, binocular rivalry produces its normal slow alternation of perceptions even if the competing stimuli are switched rapidly—several times per second—between the two eyes. If rivalry were merely a question of which eye the brain is paying attention to, the rivalry phenomenon would vanish when stimuli are switched quickly in this way. (The viewer would see, rather, a rapid alternation of the stimuli.) The observed persistence of slowly changing rivalrous perceptions when stimuli are switched strongly suggests that rivalry occurs because alternate stimulus representations compete in the visual pathway. Binocular rivalry thus affords an opportunity to study how the visual system decides what we see even when both eyes see (almost) the same thing.

A PERCEPTUAL PUZZLE

What do these findings reveal about visual awareness? First, they show that we are unaware of a great deal of activity in our brains. We have long known that we are mostly unaware of the activity in the brain that maintains the body in a stable state—one of its evolutionarily most ancient tasks. Our experiments show that we are also unaware of much of the neural activity that generates—at least in part—our conscious experiences.

We can say this because many neurons in our brains respond to stimuli that we are not conscious of. Only a tiny fraction of neurons seem to be plausible candidates for what physiologists call the "neural correlate" of conscious perception—that is, they respond in a manner that reliably reflects perception.

We can say more. The relatively few neurons whose behavior reflects perception are distributed over the entire visual pathway, rather than being part of a single area in the brain. Even though the ITC clearly has many more neurons that behave this way than those in other regions do, such neurons may be found elsewhere in future experiments. Moreover, other brain regions may be responsible for any decision resulting from whatever stimulus reaches consciousness. Erik D. Lumer and his colleagues at University College London have studied that possibility using fMRI. They showed that in humans the temporal lobe is activated during the conscious experience of a stimulus, as we found in monkeys. But other regions, such as the parietal and the prefrontal cortical areas, are activated precisely at the time at which a subject reports that the stimulus changes.

Further data about the locations of and connections between neurons that correlate with conscious experience will tell us more about how the brain generates awareness. But the findings to date already strongly suggest that visual awareness cannot be thought of as the end product of such a hierarchical series of processing stages. Instead it involves the entire visual pathway as well as the frontal parietal areas, which are involved in higher cognitive processing. The activity of a significant minority of neurons reflects what is consciously seen even in the lowest levels we looked at, V1 and V2; it is only the proportion of active neurons that increases at higher levels in the pathway.

It is not clear whether the activity of neurons in the very early areas is determined by their connections with other neurons in those areas or is the result of top-down, "feedback" connections emanating from the temporal or parietal lobes. Visual information flows from higher levels down to the lower ones as well as in the opposite direction. Theoretical studies indicate that systems with this kind of feedback can exhibit complicated patterns of behavior, including multiple stable states. Different stable states maintained by top-down feedback may correspond to different states of visual consciousness.

One important question is whether the activity of any of the neurons we have identified truly determines an animal's conscious perception. It is, after all, conceivable that these neurons are merely under the control of some other unknown part of the brain that actually determines conscious experience.

Elegant experiments conducted by William T. Newsome and his colleagues at Stanford University suggest that in area MT/V5, at least, neuronal activity can indeed determine directly what a monkey perceives. Newsome first identified neurons that selectively respond to a stimulus moving in a particular direction, then artificially activated them with small electric currents. The monkeys reported perceiving motion corresponding to the artificial activation even when stimuli were not moving in the direction indicated.

It will be interesting to see whether neurons of different types, in the ITC and possibly in lower levels, are also directly implicated in mediating consciousness. If they are, we would expect that stimulating or temporarily inactivating them would change an animal's reported perception during binocular rivalry.

A fuller account of visual awareness will also have to consider results from experiments on other cognitive processes, such as attention or what is termed working memory. Experiments by Robert Desimone and his colleagues at the National Institute of Mental Health reveal a remarkable resemblance between the competitive interactions observed during binocular rivalry and processes implicated in attention. Desimone and his colleagues train monkeys to report when they see stimuli for which they have been given cues in advance. Here, too, many neurons respond in a way that depends on what stimulus the animal expects to see or where it expects to see it. It is of obvious interest to know whether those neurons are the same ones as those firing only when a pattern reaches awareness during binocular rivalry.

The picture of the brain that starts to emerge from these studies is of a system whose processes create states of consciousness in response not only to sensory inputs but also to internal signals representing expectations based on past experiences. In principle, scientists should be able to trace the networks that support these interactions. The task is huge, but our success in identifying neurons that reflect consciousness is a good start.

MORE TO EXPLORE

Blake, Randolph, and Nikos K. Logothetis. 2002. Visual competition. *Nature Reviews Neuroscience* 3 (1): 13–21.

Crick, Francis. 1994. *The astonishing hypothesis: The scientific search for the soul.* Scribner's.

Hubel, David H. 1995. *Eye, brain, and vision.* Scientific American Library, no. 22. W. H. Freeman.

Milner, A. David, and Melvyn A. Goodale. 1996. *The visual brain in action.* Oxford University Press.

Zeki, Semir. 1993. *A vision of the brain.* Blackwell Scientific Publications.

NIKOS K. LOGOTHETIS is director of the physiology of cognitive processes department at the Max Planck Institute for Biological Cybernetics in Tübingen, Germany. He received his Ph.D. in human neurobiology in 1984 from Ludwig-Maximillians University in Munich. Since 1992 he has been adjunct professor of neurobiology at the Salk Institute in San Diego; since 1995, adjunct professor of ophthalmology at the Baylor College of Medicine; and since 2002, visiting professor of the brain and cognitive sciences department and the McGovern Center at the Massachusetts Institute of Technology. His recent work includes the application of functional imaging techniques to monkeys and the measurement of how the functional magnetic resonance imaging signal relates to neural activity.
Originally published in *The Hidden Mind,* special edition of *Scientific American,* August 2002.

8

Rethinking the 'Lesser Brain'

Long thought to be solely the brain's coordinator of body movement, the cerebellum is now known to be active during a wide variety of cognitive and perceptual activities

JAMES M. BOWER AND LAWRENCE M. PARSONS

"In the back of our skulls, perched upon the brain stem under the overarching mantle of the great hemispheres of the cerebrum, is a baseball-sized, bean-shaped lump of gray and white brain tissue. This is the cerebellum, the 'lesser brain.'"

So began, somewhat modestly, the article that in 1958 introduced the cerebellum to the readers of *Scientific American*. Written by Ray S. Snider of Northwestern University, the introduction continued, "In contrast to the cerebrum, where men have sought and found the centers of so many vital mental activities, the cerebellum remains a region of subtle and tantalizing mystery, its function hidden from investigators." But by the time the second *Scientific American* article on the cerebellum appeared 17 years later, author Rodolfo R. Llinás (currently at New York University Medical Center) confidently stated, "There is no longer any doubt that the cerebellum is a central control point for the organization of movement."

Recently, however, the cerebellum's function has again become a subject of debate. In particular, cognitive neuroscientists using powerful new tools of brain imaging have found that the human cerebellum is active during a wide range of activities that are not directly related to movement. Sophisticated cognitive studies have also revealed that damage to specific areas of the cerebellum can cause unanticipated impairments in nonmotor processes, especially in how quickly and accurately people perceive sensory information. Other findings indicate that the cerebellum may play important roles in short-term memory, attention, impulse control, emotion, higher cognition, the ability to schedule and

LARGER THAN YOU'D THINK

Flattening the outer layer, or cortex, of the two human cerebral hemispheres and the cerebellum illustrates that the cerebellum has roughly the same surface area as a single cerebral hemisphere, even though when folded it takes up much less space. The size and complexity of the cerebellum indicate that it must play a crucial function.

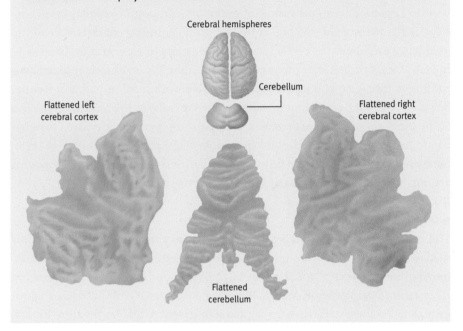

Cerebral hemispheres

Cerebellum

Flattened left
cerebral cortex

Flattened right
cerebral cortex

Flattened
cerebellum

plan tasks, and possibly even in conditions such as schizophrenia and autism. Additional neurobiological experiments—both on the pattern of sensory inputs to the cerebellum and on the ways in which the cerebellum processes that information—also suggest a need to substantially revise current thinking about the function of this organ. The cerebellum has once again become an area of "tantalizing mystery."

In retrospect, it is perhaps not surprising that the cerebellum acts as more than just a simple controller of movement. Its great bulk and intricate structure imply that it has a more pervasive and complex role. It is second in size only to the cerebral cortex, the wrinkled surface of the brain's two large hemispheres, which is known to be the seat of many critical brain functions. Like the human cerebral cortex, the cerebellum packs a prodigious amount of circuitry into a small space by folding in on itself numerous times. Indeed, the human cerebellum is much more folded than the cerebral cortex; in various mammals, it is the sole folded brain structure. Flattening out the human cerebellum yields a sheet with an average area of 1,128 square centimeters—slightly larger than a record album cover. That is more than half the 1,900 square centimeters of the surface area of the two cerebral cortices added together.

The cerebellum clearly has an important job, because it has persisted—and become larger—during the course of evolution. Although biologists often consider the growth of the cerebral cortex to be the defining characteristic of human brain evolution, the cerebellum has also enlarged significantly, increasing in size at least three times during the past million years of human history, according to fossilized skulls. Perhaps the cerebellum's most remarkable feature, however, is that it contains more individual nerve cells, or neurons, than the rest of the brain combined. Furthermore, the way those neurons are wired together has remained essentially constant over more than 400 million years of vertebrate evolution (see box, page 94). Thus, a shark's cerebellum has neurons that are organized into circuits nearly identical to those of a person's.

MORE THAN MOVEMENT

The hypothesis that the cerebellum controls movement was first proposed by medical physiologists in the middle of the 19th century, who observed that removing the cerebellum could result in immediate difficulties in coordinating movement. During World War I, English neurologist Gordon Holmes added great detail to these early findings by going from tent to tent on the front lines of battle and documenting the lack of motor coordination in soldiers who had suffered gunshot or shrapnel wounds to the cerebellum.

In the past 15 years, however, more refined testing techniques have made the story more complicated. In 1989 Richard B. Ivry and Steven W. Keele of the University of Oregon observed that patients with cerebellar injuries cannot accurately judge the duration of a particular sound or the amount of time that elapses between two sounds. In the early 1990s researchers led by Julie A. Fiez of Washington University observed that patients with damage to the cerebellum were more error-prone than others in performing certain verbal tasks. One such individual, for instance, required additional time to think of an appropriate verb, such as "to shave," when shown a picture of a razor, for example. He came up with a descriptor such as "sharp" more readily.

In more recent studies, the two of us demonstrated that patients who have neurodegenerative diseases that specifically shrink the cerebellum are often less accurate than others in judging fine differences between the pitch of two tones. Similarly, Peter Thier of the University of Tübingen in Germany and his co-workers found that people with damage to, or shrinkage of, part or all of the cerebellum are prone to make errors in tests in which they are asked to detect the presence, speed and direction of moving patterns. In addition, Hermann Ackermann and his collaborators, also at Tübingen, observed that patients with degenerated cerebellums are less able than healthy persons to discriminate between the similar-sounding words "rabbit" and "rapid."

The impairments experienced by people with cerebellar damage can extend beyond language, vision and hearing. Jeremy D. Schmahmann of Massachusetts General Hospital reported that adult and child cerebellar patients have difficulty modulating their emotions: they either fail to react to or overreact to a stimulus that elicits a more moderate response from most people. Other researchers have demonstrated that adults with cerebellar damage show delays and tend to make mistakes in spatial reasoning tests, such as determining whether the shapes of objects seen from different views match. Some scientists have also tied the cerebellum to dyslexia. Rod I. Nicolson and his colleagues at the University of Sheffield in England, for instance, found that people with dyslexia and those with cerebellar damage have similar deficits in learning ability and that dyslexics have reduced cerebellar activity during certain tasks.

Other recent studies suggest that the cerebellum might be involved in working memory, attention, mental functions such as planning and scheduling, and impulse control. In 1992 Jordan Grafman and his co-workers at the National Institutes of Health observed, for instance, that individuals with an atrophied

OVERVIEW/THE CEREBELLUM
- The cerebellum sits at the base of the brain and has complex neural circuitry that has remained virtually the same throughout the evolution of animals with backbones.
- The traditional notion that the cerebellum controls movement is being questioned by studies indicating that it is active during a wide variety of tasks. The cerebellum may be more involved in coordinating sensory input than in motor output.
- Removing the cerebellum from young individuals often causes few obvious behavioral difficulties, suggesting that the rest of the brain can learn to function without a cerebellum.

HOW THE CEREBELLUM IS WIRED

The basic features of cerebellar circuitry have been known since the seminal work of Spanish neuroanatomist Santiago Ramón y Cajal in the late 1800s. The central neuron is the Purkinje cell, named for Czech physiologist Johannes E. Purkinje, who identified it in 1837. The Purkinje cell provides the sole output of the cerebellar cortex and is one of the largest neurons in the nervous system, receiving an extraordinary 150,000 to 200,000 inputs (synapses)—an order of magnitude more than any single neuron in the cerebral cortex. These inputs spring principally from one of the smallest vertebrate neurons, the cerebellar granule cell. Granule cells are packed together at a density of six million per square millimeter, making them the most numerous type of neuron in the brain. The axon, or main trunk line carrying the outgoing signal, of every granule cell rises vertically out of the granule cell layer, making multiple inputs with its overlying Purkinje cell. The axon then splits into two segments that stretch away in opposite direc-

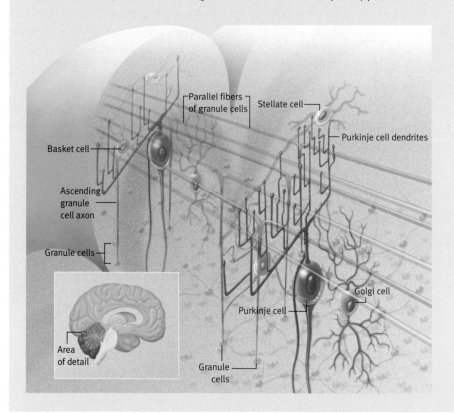

tions. These segments align into parallel fibers that run through the arms, or dendrites, of a Purkinje cell like wires through an electrical pole, providing a single input to many hundreds of Purkinje cells. Granule cells also communicate with three other types of neurons—the stellate, basket and Golgi cells—which help to modulate the signals emitted by both the granule and Purkinje cells. This basic pattern occurs in every cerebellum, indicating that it must be integral to its function. —J.M.B. and L.M.P.

cerebellum had trouble with the planning and scheduling of steps required to solve the Tower of Hanoi problem, in which rings of different sizes must be arranged on a series of pegs according to specific rules. Two independent neuroimaging studies conducted in 1997 reported that the cerebellum of healthy volunteers became active when they were asked to recall a list of letters they had heard recited moments earlier or when they were instructed to search a pattern for a specific image. In 2002 a brain imaging study by Xavier Castellanos, Judith L. Rapoport and their colleagues at the National Institute of Mental Health found that the cerebellum of children with attention-deficit hyperactivity disorder—which is characterized by a lack of impulse control—is reduced in size. Finally, brain imaging studies of healthy people and animals indicate that the cerebellum is normally active during sensory processes such as hearing, smell, thirst, need for food or air, awareness of body movement, and perception of pain.

TOUCHING, FEELING

We are among the researchers who have become convinced that the traditional motor-control theory of cerebellar function is inadequate to account for the new data. We came to this conclusion initially by intensively studying cerebellar regions that are active during touch. One of us (Bower) began such work while a graduate student more than 20 years ago in the laboratory of Wallace I. Welker at the University of Wisconsin-Madison. These investigations used a technique called micro-mapping to record the electrical activity of small patches of neurons in the brains of rats as they were touched lightly on various parts of their bodies.

Such tactile stimuli evoked activity across a large area of the cerebellum. What is more, the map appeared fractured, with neighboring areas of the cerebellum often receiving inputs from disparate regions of the body and with areas that were far apart being represented next to one another in the cerebellum. Such a mapping plan is very different from that occurring in the cerebral cortex, where the spatial relations between areas of the body surface are retained in the cortical regions that respond to and send signals to those areas.

Although the fractured nature of the cerebellar maps is unusual, an even more surprising finding was that the rat cerebellum receives input primarily from the face of the animal. This was initially confusing because Snider had shown earlier that most of the tactile region of the cat cerebellum receives input from the forepaws and that the bulk of that region in monkeys is active when the fingers are touched.

A SENSORY COORDINATOR

Given the differences in the regions of the body surface represented in the cerebellum of various types of animals, the question became: How is the mouth of a rat like the forepaws of a cat or the fingers of a monkey? The conclusion from the Wisconsin studies appeared to be that each structure is used by each animal to learn about its environment through touch. Anyone with a cat knows how much mischief its paws can cause, and anyone familiar with children recognizes how actively—and sometimes painfully—little fingers are used to gain information about the youngsters' immediate world. But rats tend to get into trouble using their mouths. The fractured structure of the touch maps in the cerebellum supported the idea that the region was somehow comparing the sensory data coming from the multiple body parts used by each animal to explore its world. These maps seemed to be organized according to the use of the body parts rather than on their absolute proximity on the body surface. The idea that the rat cerebellum was somehow comparing the sensory information coming from different parts of the face was further supported by models and experiments examining how the cerebellum responded to these inputs. What emerged was a new hypothesis of cerebellar function suggesting that the cerebellum was specifically involved in coordinating the brain's acquisition of sensory data.

Although proposing novel ideas of brain function is easy, having the ideas accepted in a field that had decided in the 1850s that the cerebellum was a motor structure is a different story. In this case, the task was made even harder by the fact that there is clearly an extremely close linkage between sensory and motor systems in the brain, especially those involving touch. To make sure that we were seeing the effects of only sensory—and not motor—activity, we needed to study people, who, unlike rats, can follow explicit instructions about when to move and when not to. It was at this point that the partnership between the two of us began. In collaboration with Peter T. Fox of the University of Texas Health Science Center in San Antonio, we designed a neuroimaging study to compare the amount of cerebellar activity induced when volunteers were instructed to use their fingers for a sensory touch discrimination task or told just to pick up and drop small objects. Under all previous theories of cerebellar function, the fine finger motor control required in picking up and dropping the small objects would be predicted to induce large amounts of cerebellar activity in the area associated with touch. Instead we found very little cerebellar activity in that region during the pick-up-and-drop task, whereas sensory exploration using the fingers caused an intense cerebellar response (see box, pages 98–99). This observation added support to our idea that the cerebellum is more involved in sensory than pure motor function and in particular that it is highly active during the process of acquiring sensory data.

LIFE WITHOUT A CEREBELLUM

Our sensory-acquisition hypothesis is but one of several new theories arising as a consequence of the growing evidence for cerebellar involvement in more than just motor control. In many cases, the new data have been accommodated by simply broadening existing motor theories to account for nonmotor results. Ivry, for instance, has advocated a "generalized timing" hypothesis of cerebellar function that suggests that the cerebellum controls the timing of body movements (such as coordinating changes in joint angles) to allow individuals to time the duration of sensory inputs such as sights and sounds.

Other researchers have posited that the cerebellum not only facilitates fine movement but also "smooths" the processing of information related to mood and thought. Schmahmann expressed such a view in 1991, and in 1996 Nancy C. Andreasen of the University of Iowa adapted the hypothesis to schizophrenia. She maintains that cerebellar deficits could underlie the disordered mental function characteristic of the disease. Other scientists have proposed that the regions of the cerebellum that have expanded dramatically during human evolution provide computational support for psychological tasks that can be offloaded from the cerebral cortex when it is overburdened.

As the number of conditions that involve changes in cerebellar activity has grown, researchers have attributed more and more functions to the cerebellum. But scientists must explain how a single brain structure whose neural circuitry is organized into a uniform, repetitive pattern can play such an integral role in so many disparate functions and behaviors.

What is even more confounding is that people can recover from cerebellar injury. Although total removal of the cerebellum initially disrupts movement coordination, individuals (particularly young ones) can, with sufficient time, regain normal function to a considerable degree. Such plasticity is a general

PASSIVE SENSING
No movement

ACTIVE SENSORY COMPARISION
No movement

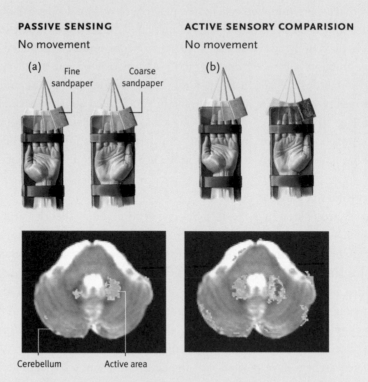

(a) Fine sandpaper   Coarse sandpaper

(b)

Cerebellum   Active area

PROBING CEREBELLAR FUNCTION

To discriminate between the cerebellum's possible roles in coordinating movement and integrating sensory input, we devised a four-part experiment. We used a technique called functional magnetic resonance imaging to reveal brain activity in the cerebellum of six healthy people while they were either sensing a stimulus on their fingers without moving them or picking up and dropping small objects. In the first scenario, we immobilized subjects' hands and rubbed pieces of sandpaper gently across their fingertips (a). Sometimes they were asked to compare the coarseness of two different types of sandpaper (b). Both were purely sensory tasks, but the latter one required subjects to discriminate between what they were feeling on each hand.

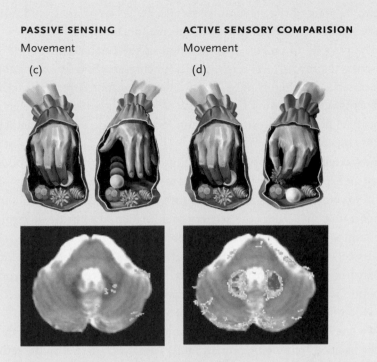

PASSIVE SENSING
Movement
(c)

ACTIVE SENSORY COMPARISION
Movement
(d)

The second scenario involved both sensory and motor aspects. A volunteer placed his or her hands into separate bags that contained small wooden balls of different shapes and textures.

In the first task (c), the person was told to randomly pick up and drop the balls, paying little heed to their shapes. In the second task (d), the individual was asked to compare the shape and feel of two balls every time he or she picked one up in each hand.

The cerebellum showed very little activity during the task that simply required picking up and dropping balls (c). In general, it was most active when the subjects were evaluating what they were sensing, either while moving (d) or still (b). These findings and others support our hypothesis that the cerebellum's main role is in processing sensory information rather than in controlling movement. —J.M.B. and L.M.P.

characteristic of the brain, but similar damage to primary sensory or motor-control regions of the cerebral cortex usually leaves animals and humans severely and permanently impaired in specific functions.

The capacity to recover from removal of the cerebellum has led some researchers to propose facetiously that its function might be to compensate for its own absence. It is highly unlikely, however, that such a large and intricate structure as the cerebellum is functionless or vestigial. Instead cerebellar function appears to permit the rest of the brain to compensate to a considerable degree for its absence.

Few cerebellar theories, including those based on motor control, have provided an explanation for this puzzling resilience. In our view, the ability of the brain to compensate for the cerebellum's absence implies a general and subtle support function. Under the sensory coordination hypothesis we favor, the cerebellum is not responsible for any particular overt behavior or psychological process. Rather it functions as a support structure for the rest of the brain. That support involves monitoring incoming sensory data and making continuous, very fine adjustments in how that information is acquired—the objective being to assure the highest possible quality of sensory input.

We predict that those adjustments take the form of extremely subtle changes in the positions of probing human fingers or rat whiskers or in the retina or the inner ear. As a support structure, the cerebellum would be expected to have some level of activity in a large number of conditions, especially those requiring careful control of incoming and perhaps remembered sensory data. Other brain systems can usually compensate for the lack of sensory data coordination through the use of alternative processing strategies if the cerebellum is damaged or removed.

Indeed, motor coordination studies suggest that people with cerebellar damage slow down and simplify their movements—reasonable strategies to compensate for a lack of high-quality sensory data. An interesting and important extension of this idea is that the continued operation of a faulty cerebellum would have more serious consequences than its complete removal. Although other brain structures can compensate for the outright lack of sensory data control, ongoing faulty control would be expected to cause continuing dysfunction in other brain regions attempting to use bad data. This type of effect might explain the recent implications for cerebellar involvement in disorders such as autism, in which patients fail to respond to incoming sensory data.

Hypotheses such as ours carry a useful reminder for future research: the presence of activity in a brain area does not necessarily mean that it is directly

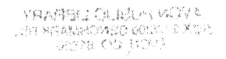

involved in a particular behavior or psychological process. Most of the machinery under the hood of a car, by analogy, is there to support the function of the engine. One could generate all kinds of hypotheses about the role of the radiator in propulsion—by correlating increased temperature to miles per hour, for instance, or by observing that a car ceases to run if its radiator is removed. But the radiator is not the engine.

If the cerebellum is primarily a support structure, then it does not contribute directly to motor coordination, memory, perception, attention, spatial reasoning or any of the many other functions recently proposed. Although this theory is one of several competing to account for the new and surprising data about the cerebellum, it is clear that how we think about this brain structure—and therefore how we conceive of the brain as a whole—is about to change.

MORE TO EXPLORE

Highstein, Stephen M., and W. Thomas Thatch, eds. 2002. *The cerebellum: Recent developments in cerebellar research*. New York Academy of Sciences.

Ivry, Richard B., and Julie A. Fiez. 2000. Cerebellar contributions to cognition and imagery. In *New cognitive neurosciences,* ed. Michael S. Gazzaniga. MIT Press.

Paulin, Michael G. 1993. The role of the cerebellum in motor control and perception. *Brain Behavior and Evolution* 41 (February): 30–50.

Schmahmann, Jeremy D., ed. 1997. *The cerebellum and cognition*. Academic Press.

JAMES M. BOWER and LAWRENCE M. PARSONS are both at the Research Imaging Center at the University of Texas Health Science Center in San Antonio, where Bower is professor of computational neurobiology and Parsons is professor of cognitive neuroscience. Bower—who is also professor at the Cajal Neuroscience Center at the University of Texas at San Antonio—is a founder of the *Journal of Computational Neuroscience* and of the international Annual Computational Neuroscience Meeting. He has also had a long-standing involvement in science education and is one of the creators of a children's educational Web site (Whyville.net). Parsons has been responsible for establishing a cognitive neuroscience program at the National Science Foundation. He is a founding member of the editorial board of the journal *Human Brain Mapping* and serves as a trustee for the International Foundation for Music Research.
Originally published in *Scientific American*, Vol. 289, No. 2, August 2003.

9

Sign Language in the Brain

How does the human brain process language? New studies of deaf signers hint at an answer

GREGORY HICKOK, URSULA BELLUGI, AND EDWARD S. KLIMA

One of the great mysteries of the human brain is how it understands and produces language. Until recently, most of the research on this subject had been based on the study of spoken languages: English, French, German and the like. Starting in the mid-19th century, scientists made large strides in identifying the regions of the brain involved in speech. For example, in 1861 French neurologist Paul Broca discovered that patients who could understand spoken language but had difficulty speaking tended to have damage to a part of the brain's left hemisphere that became known as Broca's area. And in 1874 German physician Carl Wernicke found that patients with fluent speech but severe comprehension problems typically had damage to another part of the left hemisphere, which was dubbed Wernicke's area.

Similar damage to the brain's right hemisphere only very rarely results in such language disruptions, which are called aphasias. Instead right hemisphere damage is more often associated with severe visual-spatial problems, such as the inability to copy a simple line drawing. For these reasons, the left hemisphere is often branded the verbal hemisphere and the right hemisphere the spatial hemisphere. Although this dichotomy is an oversimplification, it does capture some of the main clinical differences between individuals with damage to the left side of the brain and those with damage to the right.

But many puzzles remain. One that has been particularly hard to crack is why language sets up shop where it does. The locations of Wernicke's and Broca's areas seem to make sense: Wernicke's area, involved in speech comprehension, is located near the auditory cortex, the part of the brain that receives signals from the ears. Broca's area, involved in speech production, is located next to the part of the motor cortex that controls the muscles of the mouth and lips. But is the brain's organization for language truly based on the functions of hearing and speaking?

One way to explore this question is to study a language that uses different sensory and motor channels. Reading and writing, of course, employ vision for comprehension and hand movements for expression, but for most people these activities depend, at least in part, on brain systems involved in the use of a spoken language. The sign languages of the deaf, however, precisely fit the bill. Over the past two decades, we have examined groups of deaf signers who have suffered damage to either the right or the left hemisphere of their brains, mostly as a result of strokes. By evaluating their proficiency at understanding and producing signs, we set out to determine whether the brain regions that interpret and generate sign language are the same ones involved in spoken language. The surprising results have illuminated the workings of the human brain and may help neurologists treat the ills of their deaf patients.

THE SIGNS OF LANGUAGE

Many people mistakenly believe that sign language is just a loose collection of pantomime-like gestures thrown together willy-nilly to allow rudimentary communication. But in truth, sign languages are highly structured linguistic systems with all the grammatical complexity of spoken languages. Just as English and Italian have elaborate rules for forming words and sentences, sign languages have rules for individual signs and signed sentences. Contrary to another common misconception, there is no universal sign language. Deaf people in different countries use very different sign languages. In fact, a deaf signer who acquires a second sign language as an adult will actually sign with a foreign accent! Moreover, sign languages are not simply manual versions of the spoken languages that are used in their surrounding communities. American Sign Language (ASL) and British Sign Language, for example, are mutually incomprehensible.

Sign and spoken languages share the abstract properties of language but differ radically in their outward form. Spoken languages are encoded in acoustic-temporal changes—variations in sound over time. Sign languages, however, rely on visual-spatial changes to signal linguistic contrasts. How does this difference in form affect the neural organization of language? One might hypothesize that sign language would be supported by systems in the brain's right hemisphere because signs are visual-spatial signals. Accordingly, one could contend that the sign-language analogue of Wernicke's area in deaf signers would be near the

WHERE LANGUAGE LIVES

Two of the regions of the brain's left hemisphere that play important roles in language processing are Broca's area and Wernicke's area (there are several others). Broca's area is activated in hearing individuals when they are speaking and in deaf people when they are signing. Wernicke's area is involved in the comprehension of both speech and signs.

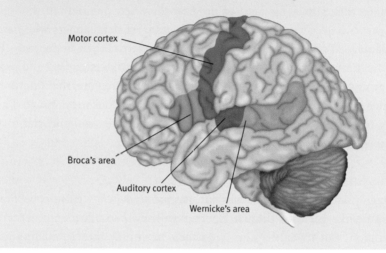

brain regions associated with visual processing and that the analogue of Broca's area would be near the motor cortex controlling hand and arm movements.

When we began to test this hypothesis in the 1980s, two fundamental questions needed to be answered: Did deaf signers with brain damage have sign-language deficits? And if so, did the deficits resemble either Wernicke's aphasia (comprehension problems and error-prone speech) or Broca's aphasia (good comprehension but difficulty in producing fluent speech)? The answer to both questions was a resounding yes. One of the first patients studied by our group signed fluently, using all the proper grammatical markers of ASL, but the message conveyed by his signing was often incoherent. An English gloss of one of his utterances reads:

> And there's one (way down at the end) [unintelligible]. The man walked over to see the (disconnected), an extension of the (earth) room. It's there for the man (can live) a roof and light with shades to (keep pulling down).

The patient's disorganized signing and apparent lack of comprehension of others' signs were very similar to the symptoms of hearing patients with Wernicke's aphasia. Another deaf patient we studied early in the research program had extreme difficulty producing signs. She had to struggle to shape and orient her hands to perform the proper movement for virtually every sign she attempted. Most of her utterances were limited to isolated signs. This was not merely a motor-control problem: when asked to copy line drawings of objects such as an elephant or a flower, she did so accurately. Also, in contrast to her severe sign-language production problems, her comprehension of sign language was excellent. This profile of language abilities parallels the symptoms of Broca's aphasia.

But where was the brain damage that caused these sign aphasias? The answer was surprising. Both patients had lesions in their left hemispheres. And the lesions were located just about where you'd expect to find them in hearing patients with similar problems. The deaf signer with comprehension difficulties had damage that included Wernicke's area, whereas the patient who had trouble making signs had damage that involved Broca's area.

These observations showed that the left hemisphere plays a crucial role in supporting sign language. But what about the right hemisphere? One would think that damage to the right hemisphere, which appears to be critically involved in many visual-spatial functions, would have a devastating effect on sign-language ability as well. But this assumption apparently is wrong. Signers with damage to the right hemisphere were fluent and accurate in their production of signs, used normal grammar and comprehended signs with ease. This held true even in patients whose nonlinguistic visual-spatial abilities had been severely compromised by their brain damage. One signer with damage to the right hemisphere, for example, could not create or copy recognizable drawings and failed to notice objects in the left part of his visual field (a condition known as hemispatial neglect). Yet he could communicate very efficiently in sign language.

Subsequent research using larger groups of deaf signers confirmed the early case studies. A study published by our team in 1996 compared the sign-language abilities of 13 left hemisphere-damaged (LHD) signers with those of 10 right hemisphere-damaged (RHD) signers. As a group, the LHD signers performed poorly across a wide range of sign-language measures: They had trouble comprehending isolated signs and signed sentences and were likely to have problems with fluency as well. They also had difficulty with picture-naming tasks and frequently made paraphasic errors—slips of the hand—in which they inadvertently substituted one sign for another or one component of a sign, such

THE BUILDING BLOCKS OF SIGN LANGUAGE

Sign languages, like spoken languages, have several kinds of linguistic structure, including phonological, morphological and syntactic levels. At the phonological level, signs are made up of a small set of components, just as spoken words are composed of a small set of consonants and vowels. The components of signs include hand shapes, the locations around the body where signs are made, the movements of the hands and arms and the orientation of the hands (for example, palm up versus palm down). In American Sign Language (ASL) the signs for "summer," "ugly" and "dry" have the same hand shape, movement and orientation but differ in location. Likewise, signs such as "train," "tape" and "chair" share hand shape, orientation and location but differ in movement.

At the morphological level, ASL has grammatical markers that systematically change the meaning of signs. Morphological markers in English include fragments like "-ed," which can be added to most verbs to indicate past tense ("walk" becomes "walked"). Whereas in English the markers are added to the beginning or end of a word, in ASL the signs are modified using distinctive spatial patterns. For example, adding a rolling movement to the sign "give" (and to most ASL verb signs) changes the sign's meaning to "give

"SUMMER"

Location of a sign is a critical element in conveying meaning. In American Sign Language, "summer" is articulated near the forehead, "ugly" near the nose and "dry" near the chin.

continuously." Signers can use different patterns to modify the verb to mean "give to all," "give to each," "give to each other" and many other variations. At the syntactic level, ASL specifies the grammatical relations among signs (that is, who is doing what to whom) in ways that do not occur in spoken languages. In English the order of the words provides the primary cue for the syntactic organization of a sentence such as "Mary criticized John." Reverse the order of the nouns, and you reverse the meaning of the sentence. Signers of ASL can use word-order cues as well, but they need not. Instead signers can point to a distinct position in space while signing a noun, thus linking the word with that position. Then the signer can move the verb sign from Mary's position to John's to mean "Mary criticized John" and in the other direction to mean the reverse.

"UGLY"

1   2

"DRY"

1   2

as hand shape, for another. In contrast, the RHD signers performed well on all these tasks. The study also showed that difficulties with sign-language fluency were not caused by more general problems in controlling voluntary hand or arm movements: patients who had trouble making signs were often capable of producing nonmeaningful hand and arm gestures.

We obtained similar results in another study, this one focusing on sign-language comprehension in 19 lifelong signers with brain lesions, 11 with damage to the left hemisphere and eight with damage to the right. The LHD group performed significantly worse than the RHD group on three tests that evaluated their understanding of single signs, simple sentences and complex sentences. The most impaired signers were those with damage to the brain's left temporal lobe, where Wernicke's area is located.

Taken together, these findings suggest that the brain's left hemisphere is dominant for sign language, just as it is for speech. The organization of the brain for language does not appear to be particularly affected by the way in which language is perceived and produced.

THE STORY GETS COMPLICATED

As we noted at the beginning of this article, the assumed left-right dichotomy of the brain—with verbal abilities concentrated in the left hemisphere and visual-spatial abilities clustered in the right—is an oversimplification. Research over the past few decades has shown that most cognitive abilities can be divided into multiple processing steps. At some levels, brain activity may be lateralized (taking place primarily in one hemisphere), whereas at others the activity may be bilateral (occurring in both).

Language ability, for instance, has many components. A hearing person must be able to perceive and produce individual speech sounds and the words they make up; otherwise, one could not distinguish "cup" from "pup." In addition, one must be able to recognize morphological additions ("walking" vs. "walked"), syntactic constructions ("the dog chased the cat" vs. "the dog was chased by the cat"), and melodic intonations ("the *White* House" vs. "the white *house*"). Finally, to conduct an extended discourse one must be able to establish and maintain a coherent connection between characters and events over the course of many sentences.

Of all these aspects of linguistic ability, the production of language is the one most sharply restricted to the brain's left hemisphere. Damage to the left hemisphere often interferes with the ability to select and assemble appropriate sounds and words when speaking. Right hemisphere damage rarely does.

One exception to the left hemisphere's monopoly on language production is the creation of a coherent discourse. Patients with right hemisphere damage may be able to construct words and sentences quite well, but they frequently ramble from one subject to the next with only a loose thread of a connection between topics.

The perception and comprehension of language appear to be less confined to the left hemisphere than language production is. Both hemispheres are capable of distinguishing individual speech sounds, and the right hemisphere seems to have a role in the comprehension of extended discourse. But deciphering the meaning of words and sentences seems to take place primarily in the left hemisphere. This may explain why language was originally considered to be the exclusive province of the left hemisphere: the most common tests for aphasia evaluated the comprehension and production of words and sentences, not longer discourses.

Nonlinguistic spatial abilities can also be broken down into components with differing patterns of lateralization. Although the most severe impairments of spatial abilities occur more commonly following damage to the right hemisphere (in both deaf and hearing populations), researchers have observed some visual-spatial deficits in LHD hearing people. The symptoms typically involve difficulties in perceiving or reproducing the local-level features of a visual stimulus—such as the details in a drawing—even though the LHD patients can correctly identify or reproduce the drawing's overall configuration. RHD hearing people tend to show the opposite pattern. Thus, it has been suggested that the left hemisphere is important for local-level spatial perception and manipulation, whereas the right hemisphere is important for global-level processes.

This more sophisticated picture of the brain raises an interesting question: Is the division of visual-spatial abilities between the two hemispheres—local level in the left, global level in the right—related to the division of sign-language abilities? Individual signs and signed sentences can be thought of as pieces of the language, whereas an extended discourse can represent how those pieces are put together. Perhaps the left hemisphere is dominant for producing and comprehending signs and signed sentences because those processes are dependent on local-level spatial abilities. And perhaps the right hemisphere is dominant for establishing and maintaining a coherent discourse in sign language because those processes are dependent on global-level spatial abilities.

We set out to test this hypothesis. Our research confirmed that many RHD signers have trouble with extended discourse: their narratives are full of tangential utterances and even confabulations—just the kind of difficulties that hearing RHD patients often have. But some RHD signers face another type of problem. Discourse in sign language has a unique spatial organization: when telling a story with many characters, the signer identifies each one using a different location. The space in front of the signer becomes a sort of virtual stage on which each character has his or her own spot. Our studies found that some RHD signers were able to stay with a topic in their discourse but failed to maintain a consistent spatial framework for the characters in their narratives.

Is either of these types of discourse problems in RHD deaf signers causally connected to deficits in their nonlinguistic spatial abilities? It would appear not. We studied one RHD signer whose spatial abilities were severely impaired yet who had no trouble signing a coherent story. Another RHD patient had only mild visual-spatial problems yet could not sustain a proper spatial framework for the characters in the narrative. Clearly, the cognitive systems in the right hemisphere that support nonlinguistic spatial abilities are different from the ones that support extended discourse.

What about deaf signers with damage to the left hemisphere? Are their sign-language aphasias caused by impairments in local-level spatial abilities? To address this issue, we asked a group of deaf signers to reproduce line drawings and hierarchical figures, which have recognizable local and global features. (An example would be the letter "D" fashioned out of a constellation of small "y"'s.) Just like hearing patients with left hemisphere damage, the LHD deaf subjects tended to reproduce the global configuration of the drawings correctly but often left out some of the details. (The RHD deaf subjects exhibited the reverse pattern, drawing pictures with lots of detail but a disorganized whole.) We found no correlation between the severity of the local-level spatial deficits in the LHD subjects and the severity of their sign-language aphasias. Contrary to all expectations, the sign-language abilities of lifelong deaf signers appear to be independent of their non-linguistic spatial skills.

It is possible that we have missed some fine distinctions in the organization of the brain for language in hearing patients and signers. Studies of patients with brain lesions are limited in their precision: to ascertain exactly which parts of the brain are involved in sign language, researchers would need to examine dozens of deaf signers with lesions in just the right places, and it would take decades to find them all. But the introduction of noninvasive brain imaging techniques—functional magnetic resonance imaging (fMRI) and positron-emission tomography (PET)—has given scientists new tools for probing the neural roots of language.

Researchers have employed these techniques to investigate the role of Broca's area in speech and sign production. Imaging results have shown that Broca's area is indeed activated in hearing patients when they are speaking and in deaf patients when they are signing. Brain imaging has also confirmed that the regions that play a role in sign-language comprehension are much the same as those involved in the understanding of spoken language. In one recent study, researchers used fMRI methods to observe the brain activity of lifelong deaf signers who were watching videotapes of sentences in ASL. The investigators found regions of activity in several parts of the left temporal lobe, including parts of Wernicke's area, and in several regions of the left frontal lobe, including Broca's area.

The study also found regions of activity in the right temporal lobe and right frontal lobe. This result has led some researchers to suggest that sign-language comprehension may be more bilaterally organized than spoken-language comprehension. But bilateral activity has also been detected in studies of hearing subjects listening to speech. More research is needed to clarify the role of the right hemisphere in sign-language processing. In any case, the studies of brain lesions make it clear that if differences exist between spoken and sign language, they are likely to be subtle and language-specific.

LESSONS FROM SIGN LANGUAGE

Sign language involves both linguistic and visual-spatial processing—two abilities that are supported by largely distinct neural systems in hearing individuals. But contrary to all expectations, the neural organization of sign language has more in common with that of spoken language than it does with the brain organization for visual-spatial processing. Why should this be the case?

The answer suggested by our line of research, as well as the work of others, is that the brain is a highly modular organ, with each module organized around a particular computational task. According to this view, the processing of visual-spatial information is not confined to a single region of the brain. Instead different neural modules process visual inputs in different ways. For example, visual inputs that carry linguistic information would be translated into a format optimized for linguistic processing, allowing the brain to access the meanings of signs, extract grammatical relations, and so on. But visual stimuli that carry a different kind of information—such as the features and contours of a drawing—would be translated into a format that is optimized for, say, carrying out motor commands to reproduce that drawing. The computational demands of these two kinds of processing tasks are very different, and thus different neural systems are involved.

When sign language is viewed in this way, it is not so surprising that comprehending and producing it appear to be completely independent of visual-spatial abilities such as copying a drawing. Although they both involve visual inputs and manual outputs, the tasks are different in fundamental ways. Consequently, we would expect them to share brain systems to some extent at the peripheral levels of processing—for instance, at the primary visual cortex that receives signals from the optic nerve—but to diverge in more central, higher-level brain systems.

The situation with spoken and sign languages is just the opposite. These two systems differ radically in their inputs and outputs but appear to involve very similar linguistic computations. We therefore expect that spoken and sign languages will share a great deal of neural territory at the more central, higher-level brain systems but diverge at the more peripheral levels of processing. At the sensory end, for example, the peripheral processing of speech occurs in the auditory cortices in both hemispheres, whereas the initial processing of signs takes place in the visual cortex. But after the first stages of processing, the signals appear to be routed to central linguistic systems that have a common neural organization in speakers and signers.

These findings may prove useful to neurologists treating deaf signers who have suffered strokes. The prognosis for the recovery of the signers' language abilities will most likely be similar to that of hearing patients with the same brain damage. Furthermore, when neurosurgeons remove brain tumors from deaf signers, they must take the same precautions to avoid damaging the language centers as they do with hearing patients.

A major challenge for future research will be to determine where the peripheral processing stages leave off and the central stages begin (or even if there is such a sharp boundary between the two). More study is also needed to understand the nature of the computations carried out at the various levels of linguistic processing. The similarities and differences between spoken and sign languages are ideally suited to answering these questions.

MORE TO EXPLORE

Emmorey, Karen. 2001. *Language, cognition, and the brain.* Lawrence Erlbaum Associates.

Hickok, G., and U. Bellugi. 2001. The signs of aphasia. In *Handbook of neuropsychology,* vol. 3, ed. R. S. Berndt. 2nd ed. Elsevier.

Hickok, G., U. Bellugi, and E. S. Klima. 1998. The neural organization of language: Evidence from sign language aphasia. *Trends in Cognitive Sciences* 2 (4): 129–136.

Klima, Edward S., and Ursula Bellugi. 1979. *The signs of language.* Harvard University Press. Repr., 1988.

Poizner, H., Edward S. Klima, and Ursula Bellugi. 1987. *What the hands reveal about the brain.* MIT Press. Reprt. Bradford Books, 1990.

GREGORY HICKOK, URSULA BELLUGI, and EDWARD S. KLIMA have worked together on sign language aphasia for a decade. Hickok is associate professor in the Department of Cognitive Sciences at the University of California, Irvine, and director of the Laboratory for Cognitive Brain Research, where he studies the functional anatomy of language. Bellugi is director of the Salk Institute's Laboratory for Cognitive Neuroscience in La Jolla, Calif., where she conducts research on language and its biological foundations. Much of her research is in collaboration with Klima, who is professor emeritus at the University of California, San Diego, and a co-director of Salk.

Originally published in *The Hidden Mind,* special edition of *Scientific American,* August 2002.

10

Hunting for Answers

A single mutation casts the death sentence of Huntington's disease. Researchers are pinning down how that mutation ruins neurons—knowledge that may suggest therapies

JUERGEN ANDRICH AND JOERG T. EPPLEN

The cups fell to the floor with a crash. Was this the alarm signal? Or was it forgetting his sister's phone number the other day, even though he calls her often? Was the telling event last weekend, when he burst into a string of curse words and tailgated the driver who had just cut him off?

Incidents that to other people may seem like simple clumsiness, forgetfulness or an overreaction brought on by stress could mean disaster for Martin, a 48-year-old shipping agent. For years, he had been observing himself and his siblings with a sharp eye. Any little slip could constitute a somber omen. But after this latest string of mishaps, he could not bear the uncertainty any longer. He went in for the blood test. Three days later what Martin had feared since childhood was confirmed as the terrible truth: he was suffering from the genetic mutation that had killed his mother, his uncle and his grandfather.

Huntington's disease was recognized as an inherited disorder more than 100 years ago, yet the mutation that causes it was not discovered until 1993. A DNA test on a blood sample was quickly devised to reveal whether a person carried the abnormal form of the gene, which leads to progressive destruction of the brain, crippling muscles and mental function. Since then, every man or woman who has had a parent or other relative with the disease has faced a vexing choice: Should he or she take the test? A positive verdict is a damnation—the disease leads to certain death, given that there is no cure. Not knowing can be easier; most people do not begin to exhibit symptoms until they are middle-aged, and the progression can be very gradual. Yet nagging suspicion can creep into every corner of life, as it did for Martin.

Of course, a cure, or even treatment that could slow the disease, would ease the tension greatly and extend life for the 30,000 Americans who have been diagnosed with Huntington's. Researchers are pinning down just how the genetic mutation ruins cellular mechanisms inside neurons, knowledge that might help point the way to therapies that have thus far proved elusive.

LETHAL KNOWLEDGE

The Huntington's test is so certain because the disease is caused by a single gene—the *huntingtin* gene on chromosome 4 (the name of the gene is spelled differently than that of the illness). Typically this gene contains several occurrences of a set of DNA building blocks: cytosine, adenine and guanine, abbreviated as CAG. This set drives the production of the huntingtin protein. The more often the CAG sequence comes up in the gene, the more glutamine—an amino acid—there is in the huntingtin protein. In healthy genes the CAG sequence may appear up to 28 times. But if it occurs more than 35 to 40 times, the glutamine chain in the huntingtin protein becomes too long and causes trouble (see box, page 118). The larger the number of CAG sets, the longer the chain, and the earlier and more severe the disease.

For Martin, the genetic test confirmed his grim suspicions. But he decided to see what he could do and went for counseling to the North Rhine-Westphalia Huntington Center at Ruhr University in Bochum, Germany. The center serves more than 600 Huntington's patients and their families. "I accept my fate as a fact," Martin told his counselors, then asked, "What can I do now that will help me later?" The team discussed the various ways the disease could play out.

The genetic mutation that overproduces the CAG sequence is inherited from a single parent. A child, therefore, has a 50 percent chance of getting the disease-causing form of the gene if either parent has it. The therapists in Bochum began by tracing the inheritance pattern of Martin's family. "My grandmother stepped in front of a train, and I think she knew what she was doing," Martin noted, insinuating that she knew she was doomed. "And Grandma's father became very strange when he got old." Martin's mother also clearly suffered from Huntington's disease, even though during her lifetime there was no definitive genetic diagnosis.

Martin and his sister Susanne finally dared to take the test because they could no longer stand the uncertainty—and because they wanted to plan the rest of their lives and careers. Susanne had the defective gene, too. Their siblings chose not to find out about their chances of impending death.

Today, two years after his weekend of dropping cups and forgetting phone numbers, Martin is showing the first verifiable symptoms: sudden twitches in his arms and legs. Susanne has been problem-free, yet she cannot help but wonder if certain harmless behaviors she never noticed before about herself are harbingers of difficulties to come. Disease symptoms typically begin when carriers are 35 to 45 years old, but even among close relatives the onset and course can differ significantly. No one wanted to think that Martin's 10-year-old son could already be falling victim—and the pediatrician treating him wanted to believe that the boy's pain, muscle weakness and subtle coordination problems were from other causes. But after six years of ailments and Martin's diagnosis, the boy was tested. Indeed, he had childhood Huntington's, a rarity brought on by an extremely elongated huntingtin gene.

INEXORABLE PROGRESSION

Huntington's disease had been known for centuries before it was given its definitive name. In the Middle Ages, victims of what was then called "the dance" made pilgrimages to Ulm, Germany, to pray in the chapel of Saint Vitus, leading to the ailment's name, Saint Vitus' dance. The first to recognize the condition as an inherited disease was the young American neurologist George Huntington in 1872. Together with his father, he had tracked cases in a family on Long Island, outside New York City, and was able to differentiate them from chorea minor, caused by a streptococcal infection that has similar symptoms. Today about one in 10,000 people in the U.S. suffers from Huntington's.

The symptoms that gave the disease its original name are the "dancing" movements—the exaggerated motions of the limbs that are its most frequent and striking effects. In the beginning, patients try to disguise these jerks and twitches as shrugs or try to translate them into deliberate motions such as stretches. But little by little their muscles go out of control. They are beset by sudden grimaces, and speaking and swallowing become increasingly difficult. In later stages, movements are slower; increased muscle contraction leads to painful contortions of the limbs that can last minutes or hours.

Characteristic mental symptoms often appear decades before the physical problems. The disease can cause repeated outbreaks of moodiness, yet patients who receive a positive test result can also fall victim to emotional swings driven by their knowledge of impending destruction. Relatives often notice personality changes—patients may become paranoid, tyrannize those around them with unfounded jealousy or react extraordinarily aggressively to trivial disagreements. As the disease advances, they may obsess about minor issues for days or weeks, burdening their families and destroying their social connections. Patients' cognitive abilities also wane; their memories deteriorate, and they find it increasingly difficult to concentrate. The problems progress to severe dementia and complete helplessness. Even early in the disease, the mounting mental breakdown can have catastrophic effects on a person's personal and professional lives, and suicide attempts are not unusual.

COMMUNICATION BREAKDOWN

Researchers who have been trying to better define the mechanisms that cause Huntington's muscular and mental challenges had to start by figuring out why the disease strikes at such varied ages. They have found that along with the huntingtin gene mutation, other inherited factors play an important role. For example, there can be great variation in how readily neuron receptors in the brain bind to glutamate, which serves as a messenger molecule that facilitates the transmission of information between neurons. The type of receptor protein a patient has helps determine how soon the disease will take hold.

Despite its rarity, Huntington's has become a focal point for defining how neurodegenerative diseases in general—including Parkinson's and Alzheimer's—harm the brain. In the 13 years since the huntingtin gene was discovered, scientists have learned much about the mechanisms leading to the destruction of neurons. Because the huntingtin protein is the sole cause of deterioration in the case of Huntington's disease, it provides a path for investigation that is uncluttered by other complicating factors.

Huntingtin is not a "bad" protein per se. It appears to play a central role in embryonic development among mammals. But as humans with the mutated gene age, the overelongated protein apparently binds to other proteins vital to cellular survival, compromising their function.

The proteins affected include transcription regulators—proteins that ensure the accurate reading of genetic information. If huntingtin proteins bind to a transcription regulator within a cell, its genetic activity is disturbed and the cell's control of protein synthesis breaks down. Some of the proteins that cannot be synthesized are responsible for removing neurotransmitters (messenger molecules) such as glutamate from the synapses—the gaps between neurons across which communication occurs. If too much glutamate remains in the synapse, adjacent neurons are continually excited; the overactivity eventually damages the cells. This phenomenon, known as excitotoxicity, has been demonstrated in lab animals.

There is also increasing evidence that huntingtin is more involved than previously thought in neuronal processes. A protein known as huntingtin interacting

LETHAL REPEATS

The huntingtin gene lies on chromosome 4. The gene contains a number of DNA building blocks represented by the sequence cytosine-adenine-guanine (CAG), which direct the production of the huntingtin protein. If the CAG sequence appears 35 to 40 times or more on the gene, the resulting protein will contain glutamine molecules that are too long, slowly killing neurons and causing Huntington's disease.

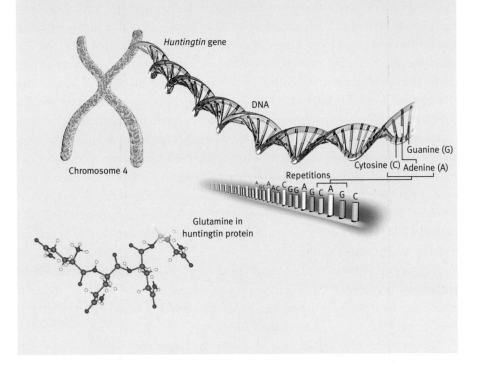

protein 1 (HIP-1) aids in both the secretion and reuptake of messenger molecules within neuronal cells. The elongated huntingtin protein chain cannot correctly bind to HIP-1, causing a cascade of enzyme reactions that initiate apoptosis: programmed cell death. The neurons are driven to kill themselves.

Another theory is based on the observation that the elongated gene causes the huntingtin protein to misfold. The misfolded protein then disrupts a neuron's metabolism. Evidence indicates that several steps in the respiration chain in the mitochondria—the cell's "power plants"—no longer function correctly. Such a resulting shortage of energy would eventually kill the cell.

SEARCH FOR A CURE

So far investigators have come up with few ideas for altering such fatal cellular disruptions. Various drug treatments have affected only the symptoms of Huntington's. For example, some neurologists prescribe neuroleptics to deal with muscle contractions. These drugs can have the unfortunate side effect of limiting the patients' mobility. Other doctors may treat their patients' psychological problems with antidepressants, sedatives or antipsychotic neuro-leptics. Unfortunately, there is still no treatment for the loss of mental abilities—a prognosis that frightened Martin more than the impending muscular challenges.

Several labs around the globe are seeking drugs that could delay or even stop the destruction of neurons. One set of substances is the glutamate antagonists, which modulate the secretion of glutamate. One compound, riluzole, has already proved effective against another serious, rapidly progressing disease of the nervous system—amyotrophic lateral sclerosis, or Lou Gehrig's disease. The drug is currently in a Europe-wide clinical test involving 450 Huntington's patients.

Scientists also have hopes for minocycline, an antibiotic. In 2003 Robert M. Friedlander of Harvard Medical School showed that in mice with Huntington's, minocycline could inhibit the action of the enzymes that set off neuronal suicide.

Other substances could perhaps block the clumping of misfolded huntingtin proteins, which aggravates neuron death. In 2004 researchers at the RIKEN Brain Science Institute in Japan inhibited the aggregation of the proteins with trehalose, a sugar made by various desert plants. Blocking the clumping in mice delayed the disease's onset.

Physicians are exploring substances produced in the body, too, such as co-enzyme Q and creatine. Coenzyme Q, ubiquitous in humans, is an antioxidant that captures oxygen free radicals and could limit damage by huntingtin proteins. Creatine, produced in the liver and kidneys, could improve energy storage in muscle and brain cells. In both cases, animal experiments have been successful, but there is no convincing evidence for efficacy in humans yet.

Researchers are testing gene therapies as well. In 2005 Beverly L. Davidson and Scott Q. Harper of the University of Iowa inhibited the action of the mutated huntingtin gene in mice. The team injected the animals' brains with RNA fragments that precisely mimicked the genetic instructions for the mutated huntingtin protein and thereby blocked its synthesis. The rodents produced less of the illness-inducing huntingtin protein.

Stem cells could provide aid. In 2000 Anne-Catherine Bachoud-Lévi of the Henri Mondor University Hospital in Creteil, France, implanted neuronal stem cells from aborted fetuses into the brains of Huntington's patients to see if they might take the place of destroyed cells. Two years later Robert A. Hauser of the University of South Florida made similar attempts. In both trials, some patients responded to the therapy, but others suffered from cerebral hemorrhages, causing their condition to worsen. All the patients had to be given additional drugs to block immune reactions to the foreign cells.

There is nothing, today, that Martin can take to block the relentless progress of his disease. But never before have so many new approaches for treatment been under investigation. Scientists from Europe and the U.S. are preparing for larger studies, and many victims, as well as others who have not had the blood test but are at risk because of their family histories, are ready to take part. With their help, hope remains for an eventual solution for patients such as Martin.

MORE TO EXPLORE

Harper, S. Q., et al. 2005. RNA interference improves motor and neuropathological abnormalities in a Huntington's disease mouse model. *Proceedings of the National Academy of Sciences USA* 102 (16): 5820–5825.

Huntington's disease: Hope through research. Online booklet with comprehensive information. Read or download at www.ninds.nih.gov/disorders/huntington/detail_huntington.htm.

Huntington's Disease Society of America. www.hdsa.org.

JUERGEN ANDRICH and JOERG T. EPPLEN are researchers at Ruhr University's North Rhine–Westphalia Huntington Center in Bochum, Germany.
Originally published in *Scientific American Mind*, Vol. 17, No. 2, April/May 2006.

11

Brain, Repair Yourself

How do you fix a broken brain? The answers may literally lie within our heads. The same approaches might also boost the power of an already healthy brain

FRED H. GAGE

For most of its 100-year history, neuroscience has embraced a central dogma: a mature adult's brain remains a stable, unchanging, computerlike machine with fixed memory and processing power. You can lose brain cells, the story has gone, but you certainly cannot gain new ones. How could it be otherwise? If the brain were capable of structural change, how could we remember anything? For that matter, how could we maintain a constant self-identity?

Although the skin, liver, heart, kidneys, lungs and blood can all generate new cells to replace damaged ones, at least to a limited extent, until recently scientists thought that such regenerative capacity did not extend to the central nervous system, which consists of the brain and spinal cord. Accordingly, neurologists had only one counsel for patients: "Try not to damage your brain, because there is no way to fix it."

Within the past five years, however, neuroscientists have discovered that the brain does indeed change throughout life—and that such revision is a good thing. The new cells and connections that we and others have documented may provide the extra capacity the brain requires for the variety of challenges that individuals face throughout life. Such plasticity offers a possible mechanism through which the brain might be induced to repair itself after injury or disease. It might even open the prospect of enhancing an already healthy brain's power to think and ability to feel.

Neuroscientists, of course, have tried to come up with fixes for brain injury or brain disorders for decades. Such treatment strategies have primarily involved replacing diminished neurotransmitters, the chemicals that convey messages between nerve cells (neurons). In Parkinson's disease, for instance, a patient's

OVERVIEW/NEW ADULT NERVE CELLS
- Naturally occurring growth factors in the adult human brain can spur the production of new nerve cells in some instances.
- The growth factors—or more easily administered drugs that prompt their production—might be useful as therapies for various brain disorders and for brain or spinal cord injuries.
- The factors could potentially be tested to enhance normal brain function, but questions remain about whether the strategy would work.

brain loses the ability to make the neurotransmitter dopamine because the cells that manufacture it die. A chemical relative of dopamine, L-dopa, can temporarily ameliorate the symptoms of the disease, but it is not a cure. Neuroscientists have also attempted to implant brain tissue from aborted fetuses to replace the neurons that perish in Parkinson's disease—and in other neurological disorders such as Huntington's and spinal cord injury—with modest success. Lately, some have turned to neurons derived from embryonic stem cells, which under the right conditions can be coaxed in laboratory dishes to give rise to all the cell types of the brain.

Although stem cell transplants have many advantages, switching on the innate capacity of the adult nervous system to repair itself would be much more straightforward. The ultimate vision is that physicians would be able to deliver drugs that would stimulate the brain to replace its own cells—and thereby rebuild its damaged circuits.

NEWBORN NERVE CELLS

Many investigators are now pursuing exactly that vision. The hope that repair might be feasible stems from a series of exciting discoveries made starting about 40 years ago. Researchers first demonstrated that the central nervous systems of mammals contain some innate regenerative properties in the 1960s and 1970s, when several groups showed that the axons, or main branches, of neurons in the adult brain and spinal cord can regrow to some extent after injury. Others (including my colleagues and me) subsequently revealed the birth of new neurons, a phenomenon called neurogenesis, in the brains of adult birds, nonhuman primates and humans (see "New Nerve Cells for the Adult Brain," by Gerd Kempermann and Fred H. Gage, *Scientific American*, May 1999).

Shortly thereafter scientists began to wonder why, if it can produce new neurons, the central nervous system fails to repair itself more reliably and completely in the wake of disease or injury. The answer lies in understanding how—and

HOW THE BRAIN MAKES NEW NEURONS

Neural stem cells are the fount of new cells in the brain. They divide periodically in two main areas: the ventricles (purple, inset), which contain cerebrospinal fluid to nourish the central nervous system, and the hippocampus (light blue, inset), a structure crucial for learning and memory. As the neural stem cells proliferate (cell pathways below), they give rise to other neural stem cells and to neural precursors that can grow up to be either neurons or support cells, which are collectively termed glial cells (astrocytes or oligodendrocytes). But these newborn neural stem cells need to move (red arrows, inset) away from their progenitors before they can differentiate. Only 50 percent, on average, migrate successfully (the others perish). In the adult brain, newborn neurons have been found in the hippocampus and in the olfactory bulbs, which process smells. Researchers hope to be able to induce the adult brain to repair itself by coaxing neural stem cells or neural precursors to divide and develop when and where they are needed. —F.H.G.

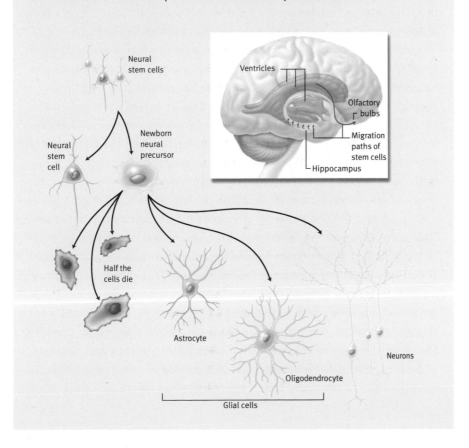

perhaps to what end—adult neurogenesis normally occurs and how the brain's natural inclination to fix itself might be amplified.

We now know that the birth of new brain cells is not a single-step process. So-called multipotent neural stem cells divide periodically in the brain, giving rise to other stem cells and to progeny that can grow up to be either neurons or support cells named glia. But to mature, these newborn cells must migrate away from the influence of the multipotent stem cells. On average, only half of them

STEM CELLS AS THERAPIES

Scientists are investigating two types of stem cells for possible use in brain-repair strategies. The first is adult neural stem cells: rare, primordial cells left over from early embryonic development that are known to occur in at least two areas of the brain and that can divide throughout life to yield new neurons as well as support cells called glia. The second is human embry-onic stem cells that have been isolated from very early human embryos, at the stage in which the embryos consist of only 100 or so cells. Such embry-onic stem cells have the potential to make any cell type in the body.

Most studies have involved observing neural stem cells while they are growing in laboratory culture dishes. Such cultured cells can multiply and be genetically marked in culture and then be transplanted back into the nervous system of an adult. In these experiments, which have so far been performed using only animals, the cells survive well and can differentiate into mature neurons in the two areas of the brain where the formation of new neurons normally occurs, the hippocampus and the olfactory bulbs. Adult neural stem cells do not readily differentiate into neurons when trans-planted into any other brain areas, although they can become glia.

The problem with adult neural stem cells is that they are still immature. Unless the adult brain into which they are transplanted is making the neces-sary signals to direct the stem cells to become a particular neural cell type, such as a hippocampal neuron, they will either die, become glial cells or merely persist as undifferentiated stem cells. The solution would be for sci-entists to determine which biochemical signals normally prompt adult neu-ral stem cells to become a particular neuronal type and then induce the cells toward that lineage in a culture dish. Once transplanted into a particular part of the brain, the cells would be expected to continue becoming that cell type, form connections with other brain cells and begin to function. —F.H.G.

make the trip; the rest die. This seemingly wasteful process mirrors that which takes place before birth and during early childhood, when more brain cells arise than are needed to form the developing brain. During that period, only those cells that form active connections with other neurons survive.

Whether the young cells that persist become neurons or glia depends on where in the brain they end up and what type of activity is occurring in that brain region at the time. It takes more than one month from when a new neuron is formed from a stem cell until it becomes fully functional and able to send and receive information. Thus, neurogenesis is a process, not an event, and one that is tightly controlled.

Neurogenesis is regulated by a variety of naturally occurring molecules called growth factors that are under intense investigation. A factor dubbed sonic hedgehog that was first discovered in insects, for example, has been shown to regulate the ability of immature neurons to proliferate. In contrast, another factor named notch and a class of molecules called the bone morphogenetic proteins appear to influence whether newborn cells in the brain become glial cells or neurons. Once young cells are committed to becoming either neurons or glial cells, other growth factors—such as brain-derived neurotrophic factor, the neurotrophins and insulin-like growth factor—play important roles in keeping the cells alive and encouraging them to mature and become functional.

WHERE THE ACTION IS

New neurons do not arise spontaneously in every part of the adult mammalian brain but appear so far to form only in fluid-filled cavities called ventricles in the forebrain and in a seahorse-shaped structure called the hippocampus that is buried deep in the brain. Researchers have shown that cells destined to become neurons travel from the ventricles to the olfactory bulbs, a pair of structures that receive input from odor-sensing cells in the nose. Although no one is sure why the olfactory bulb requires so many new neurons, we can more easily speculate why the hippocampus needs them: this structure is crucial for learning new information, so adding neurons there would presumably spur the formation of connections between new and existing neurons, increasing the brain's capacity to process and store novel information.

A handful of reports have purported to find new neurons in areas outside the hippocampus and olfactory bulb, but those results have not yet been substantiated. One reason is that the methods used to prove the existence of neurogenesis are complex and difficult to carry out. Newer, more sensitive techniques

nections. Neuroscientists speculate that these aberrant cells not only end up in the wrong place but also remain immature, contributing to the miswiring of the brain that causes seizures. Growth factor treatments for stroke, Parkinson's and other disorders might prompt neural stem cells to divide inappropriately and cause similar symptoms, so researchers must first better understand how to use the growth factors to trigger growth, the migration of new cells to specific places, or their maturation into adult cells.

In treating spinal cord injury, ALS or multiple sclerosis, the strategy may be to induce stem cells to yield a subset of glial cells called oligodendrocytes. These cells are essential for neurons to communicate with one another because they insulate the long axons between neurons, preventing the electrical signal carried by the axons from dissipating. Stem cells in the spinal cord have already been shown to have the capacity to make oligodendrocytes at low frequency. My colleagues and I—as well as other groups—have also used growth factors to induce the proliferation of oligodendrocytes in animals with spinal cord injury, with beneficial results.

A BRAIN WORKOUT

One of the most striking aspects of neurogenesis in the hippocampus is that experience can regulate the rate of cell division, the survival of newborn neurons and their ability to integrate into the existing neural circuitry. Adult mice that are moved from a rather sterile, simple cage to a larger one that has running wheels and toys, for instance, will experience a significant increase in neurogenesis. Henriette van Praag in my laboratory has found that exercising mice in a running wheel is sufficient to nearly double the number of dividing cells in the hippocampus, resulting in a robust increase in new neurons. Intriguingly, regular physical activity such as running can also lift depression in humans, perhaps by activating neurogenesis.

Once neurogenesis can be induced on demand in a controlled fashion, it could change our very conception of brain disease and injury. I imagine a time when selective drugs will be available to stimulate the appropriate steps of neurogenesis to ameliorate specific disorders. Such pharmacological therapies will be teamed with physical therapies that enhance neurogenesis and prompt particular brain regions to integrate the newly developed cells. These potential treatments offer great promise for millions of people suffering from neural diseases and spinal cord injury. The links between neurogenesis and increased mental activity and exercise also suggest that people might be able to reduce their risk of neural disease and enhance the natural repair processes in their brains by choosing a mentally challenging and physically active life.

Just as exciting is the possibility that healthy individuals might become "better than well" by stimulating their brains to grow new neurons. It is unlikely, however, that people seeking to boost their brainpower would want to have regular shots of growth factors, which cannot be taken orally and have difficulty crossing the blood-brain barrier once injected into the bloodstream. Scientists are now seeking small molecules that can be made into pills that would switch on growth factor genes in a person's brain so that the individual's brain cells make more of the factors than usual. For instance, a company named Curis, based in Cambridge, Mass., has devised small molecules that regulate the production of sonic hedgehog, a factor that plays a role in neural development. Other companies have generated similar molecules that might be made into drugs.

Another strategy that could conceivably be used to improve brain performance involves gene therapy and cell transplantation. Under such a scenario, researchers would genetically engineer cells in the laboratory to overproduce specific growth factors and then implant the cells into particular regions of a person's brain. Alternatively, scientists could insert the genes that encode the production of various growth factors into viruses that would ferry the genes into existing brain cells.

But it is not at all clear whether any of these approaches would necessarily enhance the capabilities of a normal, healthy brain. A handful of animal studies using nerve growth factor suggests that adding growth factors can actually disrupt normal brain function. It is possible that the brain requires a delicate balance and that too much of a good thing can lead to just as many problems as too little. Growth factors could induce tumors to form, and transplanted cells could potentially grow out of control, causing cancer. Such risks might be acceptable for people with diseases as dire as Huntington's, Alzheimer's or Parkinson's but might not be palatable for healthy individuals.

The best ways to augment brain function might not involve drugs or cell implants but lifestyle changes. Like many other organs, the brain responds positively to exercise, a good diet and adequate sleep, which are already known to enhance normal brain function with fewer side effects and potential problems than most of the other strategies described above. I predict that if more people knew that a proper diet, enough sleep, and exercise can increase the number of neural connections in specific regions of the brain, thereby improving memory and reasoning ability, they would take better care of themselves.

A final consideration is the environment in which we live and work. More and more experimental evidence indicates that environment can affect the wiring of the brain. This opens up vistas of possibility for architecture and suggests that future homes and offices might be designed with an eye to-

SELECTED NEURAL GROWTH FACTORS UNDER DEVELOPMENT

These factors might be used as drugs on their own, or scientists might design other drugs to stimulate or block the factors.

NAME	FUNCTION
Brain-derived neurotrophic factor [BDNF]	Keeps newborn neurons alive
Ciliary neurotrophic factor [CNTF]	Protects neurons from death
Epidermal growth factor [EGF]	Spurs stem cells in brain to divide
Fibroblast growth factor [FGF]	In low doses, supports survival of various cell types; at high doses, induces cells to proliferate
Glial cell line–derived neurotrophic factor [GDNF]	Prompts motor neurons to sprout new branches; prevents cells that perish in Parkinson's disease from dying
Glial growth factor-2 [GGF-2]	Favors production of glial [support] cells
Insulin-like growth factor [IGF]	Fosters the birth of both neurons and glial cells
Neurotrophin-3 [NT-3]	Promotes formation of oligodendrocytes [type of glial cells]

ward how they might provide an enriched environment for enhancing brain function.

More immediately, however, if science can better understand the self-healing abilities of the brain and the spinal cord, that insight could constitute one of the major achievements of our time. Neurologists of the future might be able to expand their capabilities by strategically activating the brain's own tool kit for self-repair and enhancement.

MORE TO EXPLORE

Alvarez-Buylla, Arturo, and Jose M. Garcia-Verdugo. 2002. Neurogenesis in adult subventricular zone. *Journal of Neuroscience* 22 (3): 629–634.

D'Sa, Carrol, and Ronald S. Duman. 2002. Antidepressants and neuroplasticity. *Bipolar Disorders* 4 (3): 183–194.

Kokaia, Zaal, and Olle Lindvall. 2003. Neurogenesis after ischaemic brain insults. *Current Opinion in Neurobiology* 13 (1): 127–132.

POTENTIAL DISEASE TARGETS	SOME COMPANIES INVOLVED IN RESEARCH
Depression (abandoned for amyotrophic lateral sclerosis)	Amgen, Thousand Oaks, Calif.
Huntington's disease (now testing against obesity)	Regeneron Pharmaceuticals, Tarrytown, N.Y.
Brain tumors and stroke	ImClone Systems, New York City
Brain tumors and stroke	ViaCell, Boston
Parkinson's disease and ALS	Amgen
Spinal cord injury, multiple sclerosis and schizophrenia	Acorda Therapeutics, Hawthorne, N.Y.
Multiple sclerosis, spinal cord injury, ALS and age-related dementia	Cephalon, West Chester, Pa.
Multiple sclerosis, spinal cord injury and ALS	Amgen and Regeneron Pharmaceuticals

Lie, Dieter C., et al. In press. Neurogenesis in the adult brain: New strategies for CNS diseases. In *Annual reviews of pharmacology and toxicology*.

Nottebohm, Fernando. 2002. Why are some neurons replaced in adult brains? *Journal of Neuroscience* 22 (3): 624–628.

FRED H. GAGE is Adler Professor in the Laboratory of Genetics at the Salk Institute for Biological Studies in San Diego and an adjunct professor at the University of California, San Diego. He received his Ph.D. in 1976 from Johns Hopkins University. Before joining the Salk Institute in 1994, Gage was a professor of neuroscience at UCSD. He is a fellow of the American Association for the Advancement of Science and a member of both the National Academy of Sciences and the Institute of Medicine. He served as president of the Society for Neuroscience in 2002, and his honors include the 1993 Charles A. Dana Award for Pioneering Achievements in Health and Education, the 1997 Christopher Reeve Research Medal, the 1999 Max Planck Research Prize and the 2002 MetLife Award.

Originally published in *Scientific American*, Vol. 289, No. 3, September 2003.

12

Diagnosing Disorders

Psychiatric illnesses are often hard to recognize, but genetic testing and
neuroimaging could someday be used to improve detection

STEVEN E. HYMAN

Accurate diagnosis is the cornerstone of medical care. To plan a successful treatment for a patient, a doctor must first determine the nature of the illness. In most branches of medicine, physicians can base their diagnoses on objective tests: a doctor can examine x-rays to see if a bone is broken, for example, or extract tissue samples to search for cancer cells. But for some common and serious psychiatric disorders, diagnoses are still based entirely on the patient's own report of symptoms and the doctor's observations of the patient's behavior. The human brain is so enormously complex that medical researchers have not yet been able to devise definitive tests to diagnose illnesses such as schizophrenia, autism, bipolar disorder or major depression.

Because psychiatrists must employ subjective evaluations, they face the challenge of reliability: how to ensure that two different doctors arrive at the same diagnosis for the same patient. To address this concern, the American Psychiatric Association in 1980 published the *Diagnostic and Statistical Manual of Mental Disorders*, Third Edition (widely known by the acronym *DSM-III*). Unlike earlier editions of the manual, *DSM-III* and its successor volumes (the latest one is referred to as *DSM-IV-TR*) describe what symptoms must be present—and for how long—to make a diagnosis of a particular brain disorder. Virtually all these criteria, however, are based on the patient's history and the clinical encounter. Without the ability to apply objective tests, physicians may fail to detect disorders and sometimes mistake the symptoms of one illness for those of another. Making the task more difficult is the fact that some psychiatric illnesses, such as schizophrenia, may turn out to be clusters of diseases that have similar symptoms but require different treatments.

OVERVIEW/IMPROVING DIAGNOSIS
- Because psychiatrists lack objective tests for detecting brain disorders, they sometimes fail to observe mental illness or mistake the symptoms of one disorder for those of another.
- Scientists have recently found gene variants that seem to confer susceptibility to disorders such as schizophrenia and autism. Doctors may someday be able to determine a patient's risk of developing these diseases by analyzing his or her DNA.
- In addition, advances in neuroimaging may allow physicians to look for subtle anomalies in the brain caused by mental disorders. As the technology improves, doctors could use neuroimaging to diagnose psychiatric illnesses and to track the success of therapy.

In recent years, though, advances in genetics, brain imaging and basic neuroscience have promised to change the way that brain disorders are diagnosed. By correlating variations in DNA with disease risks, researchers may someday be able to determine which small differences in a patient's genetic sequence can make that person more vulnerable to schizophrenia, autism or other illnesses. And rapid developments in neuroimaging—the noninvasive observation of a living brain—may eventually enable doctors to spot structural features or patterns of brain activity that are characteristic of certain disorders. Better diagnosis will lead to better care: after pinpointing a patient's brain disorder, a physician will be able to prescribe the treatment that is best suited to it. And earlier diagnosis could allow doctors to slow or halt the progress of a disorder before it becomes debilitating.

HISTORY OF DIAGNOSIS

The first modern attempt to identify individual psychiatric disorders was made in the 19th century by German scientist Emil Kraepelin, who distinguished two of the most severe mental illnesses: schizophrenia, which he called dementia praecox, and manic-depressive illness, which is now known as bipolar disorder. Much of his careful observational work focused on following the course of the illnesses over the lifetime of his patients. He defined schizophrenia as a disease with psychotic symptoms (such as hallucinations and delusions) that had an insidious onset—in other words, the initial symptoms may be hard to detect—and a chronic, downhill course. In contrast, manic-depressive illness was characterized by discrete episodes of illness alternating with periods of relatively healthy mental function.

In the early 20th century, however, work on psychiatric diagnosis went into eclipse as a result of the influence of the psychoanalytic theories developed by Sigmund Freud and his followers. In their conception of mental illness, symptoms arose from a failure to successfully negotiate stages in psychological development. The symptoms of each illness indicated the point in development at which the trouble arose. The psychoanalytic theory of that period did not allow for the possibility that different psychiatric illnesses might have completely different causes, let alone the modern idea that mental disorders might arise from abnormalities in brain circuits.

Diagnosis returned to a central position in psychiatry in the 1950s, though, with the discovery of drugs for treating psychiatric disorders. Researchers found that chlorpromazine (better known by one of its brand names, Thorazine) could control the psychotic symptoms of schizophrenia and that lithium salts could stabilize the moods of patients with bipolar disorder. By 1960 the first antidepressant and antianxiety drugs were introduced. It quickly became critically important to match the patient with the right treatment. The new antidepressants did not work for schizophrenia and could precipitate an episode of mania in someone with bipolar disorder. Lithium was remarkably effective for bipolar disorder but not for schizophrenia.

In the 1980s the publication of *DSM-III* and subsequent manuals enabled psychiatrists to use standardized interviews and checklists of symptoms to make their diagnoses. Although the checklist approach is imperfect, it represented an enormous advance in both clinical care and research. For example, before the advent of *DSM-III*, it appeared that schizophrenia was twice as prevalent in the U.S. as it was in Great Britain. This discrepancy turned out to be an artifact of divergent approaches to diagnosis. In fact, the prevalence of schizophrenia is about 1 percent of people worldwide. The standardization of diagnosis made it clear that mental disorders are common and quite often disabling. According to the World Health Organization's data on the global burden of disease, major depression is the leading cause of disability in the U.S. and other economically advanced nations. In aggregate, mental disorders rank second only to cardiovascular diseases in terms of their economic and social costs in those countries.

Meanwhile advances in neuroscience showed that certain neurological diseases leave unmistakable signatures on the brain. Parkinson's disease, for instance, is characterized by the death of nerve cells in the midbrain that make the neurotransmitter dopamine, a chemical that transmits signals between neurons. The definitive signs of Alzheimer's disease are deposits of an abnormal protein called amyloid and tangles of protein in the cells of the cerebral cortex, the outermost layer of the brain. (Because one needs a microscope to observe these anomalies, a conclusive diagnosis can be made only after the patient's death.) But when it comes to psychiatric illnesses such as schizophrenia and depression, the abnormalities in the brain are much more subtle and difficult to discover. For this reason, many researchers have begun to look for indicators of brain disorders in the human genome.

THE GENETICS OF DISORDER

Just as normal behavioral traits are often passed from parent to child, certain mental disorders run in families. To determine whether the resemblance is a result of genes or family environment, researchers have conducted studies comparing the risk of illness in identical twins (who share 100 percent of their DNA) to the risk in fraternal twins (who on average share 50 percent of their DNA). Another type of study, which is more cumbersome, focuses on whether an illness in offspring who were adopted early in life is more often shared with their biological relatives or their adoptive families.

Such studies reveal that genes play a substantial role in the transmission of mental disorders but that other factors must also be at work. For example, if one identical twin has schizophrenia, the risk to the other is about 50 percent. If one identical twin has autism—a developmental brain disorder characterized by impairments in communication and social interaction—the other twin has a 60 percent chance of sharing the same diagnosis. These are enormous increases over the risks for the general population (1 percent for schizophrenia, 0.2 percent for autism), but the key point here is that some twins do not develop the disorders even if they carry the same genes as their affected siblings.

Therefore, nongenetic factors must also contribute to the risk of illness. These factors may include environmental influences (such as infections or injuries to the brain early in life) and the random twists and turns of brain development. Even among identical twins growing up in exactly the same environment, it is not possible to wire a brain with 100 trillion synapses in identical fashion. For all mental disorders—and, indeed, for all normal patterns of behavior that have been studied—genes are important, but they are not equivalent to fate. Our brains, not our genes, directly regulate our behavior, and our brains are the products of genes, environment and chance operating over a lifetime.

What is more, new research indicates that the strong genetic influence on the risk of developing a disorder such as schizophrenia is not the work of a single gene. Rather, the increase in risk seems to be an aggregate effect of many genes interacting with one another and with nongenetic factors. By studying the DNA sequences of people with schizophrenia and their family members, researchers have already found several genetic variations that appear to increase susceptibility to the disorder. These variations occur in genes that code proteins involved in the transmission of signals among neurons in the brain, so it is possible that the variations disrupt that process. Similar studies have identified genetic variations that appear to increase the risk of developing major depression and bipolar disorder. Furthermore, a variation of *HOXA1*, a gene related to early brain development, seems to boost susceptibility to autism. The variant gene is present in about 20 percent of the general population but in about 40 percent of people with autism.

Although possessing the variation of *HOXA1* approximately doubles the risk of developing autism, more than 99.5 percent of people who have the variant gene do not acquire the disorder, and about 60 percent of people with autism do not possess the variant gene. As is the case for many diseases, there is not likely to be a single set of genes that are necessary and sufficient to cause either schizophrenia or autism. Instead these illnesses may arise by several pathways. This situation, called genetic complexity, seems to apply to bipolar disorder and depression as well. Each of these disorders may actually represent a group of closely related mental illnesses that share key aspects of abnormal physiology and symptoms but may differ in details large and small, including severity and responsiveness to treatment.

What are the implications for diagnosis? Imagine that variations in 10 distinct genes can boost the risk of developing a mental illness but that none of the genetic variations by itself is either necessary or sufficient to bring on the disorder (this is close to a current model for autism). Different combinations of the variant genes may confer risks of similar but not identical forms of the illness. To correlate all the possible genetic combinations with all the clinical outcomes would be an immensely complex task. But the tools for such an undertaking are already available. Thanks to technologies developed for the Human Genome Project, scientists can rapidly determine what variations are present in a person's DNA. Using gene chips—small glass slides holding arrays of thousands of reference DNA samples—researchers can also discover which genes are actively coding proteins in a given cell or tissue.

If the gene-hunting effort is successful, doctors will someday be able to analyze a patient's genetic sequence and see where it fits in the matrix of risks. The accuracy of this matrix would be greatly enhanced if physicians also had more information about environmental risk factors. In all likelihood, none of the environmental influences has an overwhelming effect on illness risk—otherwise, researchers would have probably noticed it by now—so epidemiologists will need to study large numbers of people to tease out all the small contributions. By taking both genetic and environmental factors into account, this method may be able to determine whether a person is at high risk for acquiring a particular brain disorder. High-risk patients could then receive close scrutiny in follow-up observations, and if symptoms of the disorder appear, doctors would be able to begin treatment at the earliest stages of the illness.

For patients already showing symptoms of a disorder, their genetic information would be quite useful in narrowing down the diagnostic possibilities. And as researchers learn how genetic variations can affect responses to drugs, knowing a patient's genomic profile could help a physician choose the best treatment. But there is a downside to this medical advance: in a society where people can carry their DNA sequences on a memory chip, policymakers would have to grapple with the question of who should have access to this data. Even though a genetic sequence by itself cannot definitively predict whether a person will descend into depression or psychosis, one can readily imagine how employers, educational institutions and insurance companies might use or misuse this information. Society at large will have to become far more sophisticated in its interpretation of the genetic code.

IMAGING THE BRAIN

Moving in parallel with the genomic revolution, neuroscientists have dramatically improved their ability to image the living brain noninvasively. There are three major types of neuroimaging studies. The first is morphometric analysis, which generally relies on high-resolution magnetic resonance imaging (MRI) to produce precise measurements of brain structures. The second is functional neuroimaging, which generates maps of brain activity by detecting signals that correlate with the firing of brain cells. Functional neuroimaging usually involves the application of MRI or positron-emission tomography (PET). The third type of neuroimaging, which typically employs PET, uses radioactive tracers to locate and quantify specific molecules in the brain. In research settings, imaging tools can help explain what goes wrong in the brain to produce certain mental illnesses, and these findings in turn can help define the boundaries of brain disorders. In clinical settings, neuroimaging tools may eventually play a role in diagnosis and in monitoring the effectiveness of treatment.

To be useful for psychiatric diagnosis, a test based on neuroimaging must be affordable and feasible to administer. It must also be sensitive enough to detect the inconspicuous features of a particular brain disorder and yet specific enough to rule out other conditions. Some anatomical signs of mental disorders are nonspecific: people with schizophrenia generally have enlarged cerebral

TELLTALE SIGNS IN THE BRAIN

Three-dimensional maps of the brain derived from magnetic resonance imaging reveal that one type of schizophrenia causes a characteristic pattern of tissue loss in the cerebral cortex. The maps show that the average annual reduction in the cortical gray matter of adolescent patients suffering from childhood-onset schizophrenia (right) is much greater than the loss in healthy children (left) between the ages of 13 and 18.

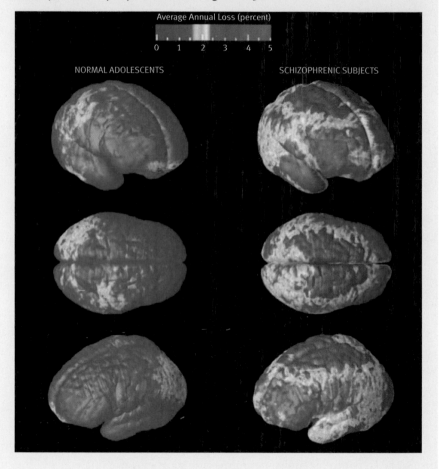

Average Annual Loss (percent)

0   1   2   3   4   5

NORMAL ADOLESCENTS   SCHIZOPHRENIC SUBJECTS

ventricles (the fluid-filled spaces deep in the brain), but this abnormality may also occur in people with alcoholism or Alzheimer's. In patients with severe, chronic depression, the hippocampus—a brain structure critically involved in memory—may be atrophied, but this anomaly has also been observed in post-traumatic stress disorder and is characteristic of the later stages of Alzheimer's.

THE SPECTRUM OF PSYCHIATRIC ILLNESS

MENTAL DISORDERS, which afflict millions of people every year, can be hard to diagnose. As the table shows, some illnesses have overlapping symptoms. Certain mood disorders, such as major depression and dysthymia, have similar symptoms but differ in severity. Among anxiety disorders, the primary distinction is the trigger that initiates fear, panic or avoidance behavior. Psychotic disorders also range from mild to severe. More definitive diagnostic methods are clearly needed.

DISORDER	COMMON SYMPTOMS	PREVALENCE (PERCENT)*
MOOD DISORDERS		
Major Depression	Characterized by episodes during which the patient feels sad or empty nearly every day; loses interest or pleasure in hobbies and activities; experiences changes in appetite, weight, energy levels or sleeping patterns; or harbors thoughts of death or suicide	5.3
Dysthymia	Similar to major depression, but the symptoms are less severe and more chronic. Sad or empty mood on most days for at least two years. Other symptoms include low self-esteem, fatigue and poor concentration	1.6
Bipolar I	Episodes of abnormally elevated or irritable mood during which the patient feels inflated self-esteem; needs less sleep; talks more than usual; or engages excessively in pleasurable but unwise activities. These manic periods may alternate with depressive episodes	1.1
Bipolar II	Depressive episodes alternate with less severe manic periods that do not markedly impair functioning or require hospitalization	0.6
ANXIETY DISORDERS		
Specific Phobia	Excessive or unreasonable fear of a specific object or situation, such as flying, heights, animals, receiving an injection or seeing blood. Exposure to the stimulus may provoke a panic attack (palpitations, sweating, trembling, shortness of breath, etc.)	
Agoraphobia	Anxiety about being in any place or situation from which escape might be difficult. Typical fears involve being alone outside the home, standing in a crowd, crossing a bridge, or traveling in a bus, train or automobile	4.9
Post-traumatic Stress Disorder	Patient persistently reexperiences a traumatic event through distressing recollections, recurring dreams or intense reactions to anything symbolizing or resembling the event	3.6
PSYCHOTIC DISORDERS		
Schizophrenia	Characterized by delusions, hallucinations, disorganized speech, inappropriate or blunted emotional responses, loss of motivation and cognitive deficits	1.3
Schizophreniform Disorder	Similar to schizophrenia, but the symptoms last for less than six months and may not be severe enough to impair social or occupational functioning	0.1

*Percent of U.S. population between ages 15 and 54 suffering from the disorder in any one-year period.

The utility of imaging for diagnosis will depend on finding abnormalities that are specific to a certain disease or perhaps to a symptom complex that may occur as a component of one or more diseases.

Furthermore, morphometric analysis of the human brain has proved to be challenging. Because the overall sizes and shapes of people's brains differ so much, researchers must employ complex computer algorithms to define normal values for various populations and compare the brains of individuals against those group norms. Moreover, the boundaries between brain structures may be very subtle. MRI atlases showing the anatomy of the normal human brain as it develops over the course of childhood and adolescence are only now becoming available.

Nevertheless, scientists have been able to use neuroimaging to shed some light on psychiatric illnesses. In 2001 teams led by Judith L. Rapoport of the National Institute of Mental Health and Paul Thompson and Arthur W. Toga of the David Geffen School of Medicine at UCLA produced an impressive study that found striking anatomical changes in the brains of adolescents with schizophrenia. The researchers focused on a relatively rare form of schizophrenia that begins in childhood. (The first signs of schizophrenia usually appear in the late teens or early 20s.) MRI scans of the brains of the affected children showed a remarkable loss of gray matter in the cerebral cortex—the brain structure responsible for higher thought—between the ages of 13 and 18 (see illustration, page 138). As the disease progressed, the loss of gray matter intensified and spread, engulfing cortical regions that support associative thinking, sensory perception and muscle movement. The anatomical abnormalities mirrored the severity of the psychotic symptoms and the impairments caused by the disease.

Such studies point the way toward a diagnostic test. It is possible that some index of measurements of cortical thickness and the size of structures known to be affected in schizophrenia (such as the hippocampus) could be used to discern whether a young person is suffering from the disorder and to chart the progress of the disease. Early detection of schizophrenia could be a great boon to treatment. Researchers are now investigating whether early intervention in schizophrenia with antipsychotic drugs and stress-management therapy can delay the onset of symptoms and reduce their severity.

Functional neuroimaging may also find significant uses in diagnosis. In Alzheimer's, loss of brain function may precede the macroscopic atrophy of brain structures. Investigators are already trying to refine the diagnosis for Alzheimer's by linking cognitive testing with functional imaging using MRI or PET. A similar strategy could possibly be applied to schizophrenia, which is characterized by failures in working memory (the ability to keep information in mind and manipulate it). It is conceivable that cognitive tests combined with functional imaging of the prefrontal cortex—a brain region that supports working memory—could contribute to the diagnosis of schizophrenia and, perhaps more important, track the success of therapy.

By combining neuroimaging with genetic studies, physicians may eventually be able to move psychiatric diagnoses out of the realm of symptom checklists and into the domain of objective medical tests. Genetic testing of patients could reveal who is at high risk for developing a disorder such as schizophrenia or depression. Doctors could then use neuroimaging on the high-risk patients to determine whether the disorder has actually set in. I do not want to sound too optimistic—the task is daunting. But the current pace of technological development augurs well for progress.

MORE TO EXPLORE

Cloninger, C. Robert. 2002. The discovery of susceptibility genes for mental disorders. *Proceedings of the National Academy of Sciences USA* 99 (21): 13365–13367. Available online at www.pnas.org/cgi/content/full/99/21/13365.

Diagnostic and statistical manual of mental disorders. 1994. 4th ed. American Psychiatric Association.

Hyman, Steven E. 2002. Neuroscience, genetics, and the future of psychiatric diagnosis. *Psychopathology* 35 (2–3): 139–144.

Thompson, Paul M., Christine Vidal, Jay N. Giedd, Peter Gochman, Jonathan Blumental, Robert Nicolson, Arthur W. Toga, and Judith L. Rapoport. 2001. Mapping adolescent brain change reveals dynamic wave of accelerated gray matter loss in very early-onset schizophrenia. *Proceedings of the National Academy of Sciences USA* 98 (20): 11650–11655. Available online at www.pnas.org/cgi/content/full/98/20/11650.

From an early age, STEVEN E. HYMAN was curious about how our brains underlie thinking, emotion and behavioral control. He studied philosophy as an undergraduate at Yale University and philosophy of science at the University of Cambridge, where he was a Mellon Fellow. After earning his M.D. at Harvard University, he received clinical training in psychiatry and scientific training in molecular neurobiology. He was the founding director of Harvard's Interfaculty Initiative in Mind, Brain and Behavior. From 1996 to 2001 he served as director of the National Institute of Mental Health, the component of the National Institutes of Health charged with generating the knowledge needed to understand, treat and prevent mental illness. Since 2001 he has been Harvard's provost and a professor of neurobiology at Harvard Medical School. Originally published in *Scientific American*, Vol. 289, No. 3, September 2003.

13

The Addicted Brain

Drug abuse produces long-term changes in the reward circuitry of the brain. Knowledge of the cellular and molecular details of these adaptations could lead to new treatments for the compulsive behaviors that underlie addiction

ERIC J. NESTLER AND ROBERT C. MALENKA

White lines on a mirror. A needle and spoon. For many users, the sight of a drug or its associated paraphernalia can elicit shudders of anticipatory pleasure. Then, with the fix, comes the real rush: the warmth, the clarity, the vision, the relief, the sensation of being at the center of the universe. For a brief period, everything feels right. But something happens after repeated exposure to drugs of abuse—whether heroin or cocaine, whiskey or speed.

The amount that once produced euphoria doesn't work as well, and users come to need a shot or a snort just to feel normal; without it, they become depressed and, often, physically ill. Then they begin to use the drug compulsively. At this point, they are addicted, losing control over their use and suffering powerful cravings even after the thrill is gone and their habit begins to harm their health, finances and personal relationships.

Neurobiologists have long known that the euphoria induced by drugs of abuse arises because all these chemicals ultimately boost the activity of the brain's reward system: a complex circuit of nerve cells, or neurons, that evolved to make us feel flush after eating or sex—things we need to do to survive and pass along our genes. At least initially, goosing this system makes us feel good and encourages us to repeat whatever activity brought us such pleasure.

But new research indicates that chronic drug use induces changes in the structure and function of the system's neurons that last for weeks, months or years after the last fix. These adaptations, perversely, dampen the pleasurable effects of a chronically abused substance yet also increase the cravings that trap the addict in a destructive spiral of escalating use and increased fallout at work

OVERVIEW/THE EVOLUTION OF ADDICTION

- Drugs of abuse—cocaine, alcohol, opiates, amphetamine—all commandeer the brain's natural reward circuitry. Stimulation of this pathway reinforces behaviors, ensuring that whatever you just did, you'll want to do again.
- Repeated exposure to these drugs induces long-lasting adaptations in the brain's chemistry and architecture, altering how individual neurons in the brain's reward pathways process information and interact with one another.
- Understanding how chronic exposure to drugs of abuse reshapes an addict's brain could lead to novel, more broadly effective ways to correct the cellular and molecular aberrations that lie at the heart of all addiction.

and at home. Improved understanding of these neural alterations should help provide better interventions for addiction, so that people who have fallen prey to habit-forming drugs can reclaim their brains and their lives.

DRUGS TO DIE FOR

The realization that various drugs of abuse ultimately lead to addiction through a common pathway emerged largely from studies of laboratory animals that began about 40 years ago. Given the opportunity, rats, mice and nonhuman primates will self-administer the same substances that humans abuse. In these experiments, the animals are connected to an intravenous line. They are then taught to press one lever to receive an infusion of drug through the IV, another lever to get a relatively uninteresting saline solution, and a third lever to request a food pellet. Within a few days, the animals are hooked: they readily self-administer cocaine, heroin, amphetamine and many other common habit-forming drugs.

What is more, they eventually display assorted behaviors of addiction. Individual animals will take drugs at the expense of normal activities such as eating and sleeping—some even to the point that they die of exhaustion or malnutrition. For the most addictive substances, such as cocaine, animals will spend most of their waking hours working to obtain more, even if it means pressing a lever hundreds of times for a single hit. And just as human addicts experience intense cravings when they encounter drug paraphernalia or places where they have scored, the animals, too, come to prefer an environment that they associ-

THE BRAIN UNDER THE INFLUENCE

CHRONIC USE of addictive substances can change the behavior of a key part of the brain's reward circuit: the pathway extending from the dopamine-producing nerve cells (neurons) of the ventral tegmental area (VTA) to dopamine-sensitive cells in the nucleus accumbens. Those changes, induced in part by the molecular actions depicted at the right and in the graph, contribute significantly to the tolerance, dependence and craving that fuel repeated drug use and lead to relapses even after long periods of abstention. The colored arrows on the brain indicate some of the pathways linking the nucleus accumbens and BTA with other regions that can help to make drug users highly sensitive to reminders of past highs, vulnerable to relapses when stressed, and unable to control their urges to seek drugs.

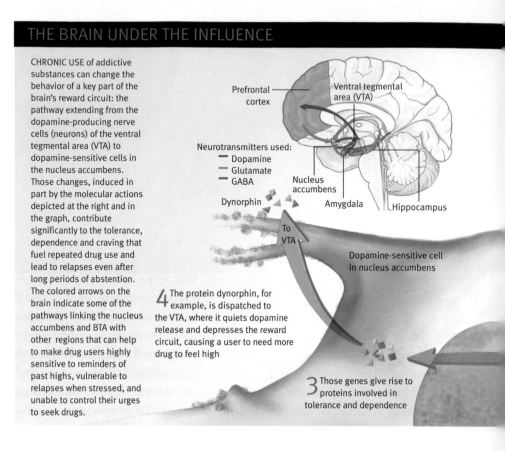

Prefrontal cortex

Ventral tegmental area (VTA)

Neurotransmitters used:
- Dopamine
- Glutamate
- GABA

Dynorphin

Nucleus accumbens

Amygdala

Hippocampus

To VTA

Dopamine-sensitive cell in nucleus accumbens

4 The protein dynorphin, for example, is dispatched to the VTA, where it quiets dopamine release and depresses the reward circuit, causing a user to need more drug to feel high

3 Those genes give rise to proteins involved in tolerance and dependence

ate with the drug—an area in the cage in which lever pressing always provides chemical compensation.

When the substance is taken away, the animals soon cease to labor for chemical satisfaction. But the pleasure is not forgotten. A rat that has remained clean—even for months—will immediately return to its bar-pressing behavior when given just a taste of cocaine or placed in a cage it associates with a drug high. And certain psychological stresses, such as a periodic, unexpected foot shock, will send rats scurrying back to drugs. These same types of stimuli— exposure to low doses of drug, drug-associated cues or stress—trigger craving and relapse in human addicts.

Using this self-administration setup and related techniques, researchers

TIMING MAKES A DIFFERENCE

Activity level

CREB   ΔFosB

Exposure to drug

1   2   3   4   5
Days   Last exposure

Dopamine-producing nerve cell of VTA

CREB: A Source of Tolerance

1 Dopamine signaling leads to increases in cyclic AMP ($cAMP$) and calcium ion (Ca^{2+}) concentrations

Dopamine

Dopamine receptor

$cAMP$ ΔFosB

Ca^{2+}

2 Those increases rapidly activate a protein called CREB. Then CREB bound to DNA activates specific genes

Nucleus

CREB

Genes activated by CREB

Delta FosB: A Key to Craving

1 Dopamine signaling also leads to the production of the protein delta FosB (ΔFosB)

2 ΔFosB represses dynorphin synthesis and activates specific genes (different from those switched on by CREB)

Dynorphin gene

X
No dynorphin

Gene activated by ΔFosB

3 The activated genes give rise to proteins involved in sensitizing responses to drugs and to reminders of past drug use

4 The protein CDK5, for example, may promote structural changes that could make nucleus accumbens neurons persistently sensitive to drugs and drug-related cues

CDK5

WHETHER a user is tolerant to a drug or, conversely, sensitized to it depends in part on the levels of active CREB and ΔFosB in nucleus accumbens cells. Initially CREB dominates, leading to tolerance and, in the drug's absence, discomfort that only more drug can cure. But CREB activity falls within days when not boosted by repeated hits. In contrast, ΔFosB concentrations stay elevated for weeks after the last drug exposure. As CREB activity declines, the dangerous long-term sensitizing effects of ΔFosB come to dominate.

mapped the regions of the brain that mediate addictive behaviors and discovered the central role of the brain's reward circuit. Drugs commandeer this circuit, stimulating its activity with a force and persistence greater than any natural reward.

A key component of the reward circuitry is the mesolimbic dopamine system: a set of nerve cells that originate in the ventral tegmental area (VTA), near the base of the brain, and send projections to target regions in the front of the brain—most notably to a structure deep beneath the frontal cortex called the nucleus accumbens. Those VTA neurons communicate by dispatching the chemical messenger (neurotransmitter) dopamine from the terminals, or tips, of their long projections to receptors on nucleus accumbens neurons. The dopamine pathway from the VTA to the nucleus accumbens is critical for addiction: animals with lesions in these brain regions no longer show interest in substances of abuse.

RHEOSTAT OF REWARD

Reward pathways are evolutionarily ancient. Even the simple, soil-dwelling worm *Caenorhabditis elegans* possesses a rudimentary version. In these worms, inactivation of four to eight key dopamine-containing neurons causes an animal to plow straight past a heap of bacteria, its favorite meal.

In mammals, the reward circuit is more complex, and it is integrated with several other brain regions that serve to color an experience with emotion and direct the individual's response to rewarding stimuli, including food, sex and social interaction. The amygdala, for instance, helps to assess whether an experience is pleasurable or aversive—and whether it should be repeated or avoided—and helps to forge connections between an experience and other cues; the hippocampus participates in recording the memories of an experience, including where and when and with whom it occurred; and the frontal regions of the cerebral cortex coordinate and process all this information and determine the ultimate behavior of the individual. The VTA-accumbens pathway, meanwhile, acts as a rheostat of reward: it "tells" the other brain centers how rewarding an activity is. The more rewarding an activity is deemed, the more likely the organism is to remember it well and repeat it.

Although most knowledge of the brain's reward circuitry has been derived from animals, brain-imaging studies conducted over the past 10 years have revealed that equivalent pathways control natural and drug rewards in humans. Using functional magnetic resonance imaging (fMRI) or positron-emission tomography (PET) scans (techniques that measure changes in blood flow associated with neuronal activity), researchers have watched the nucleus accumbens in cocaine addicts light up when they are offered a snort. When the same addicts are shown a video of someone using cocaine or a photograph of white lines on a mirror, the accumbens responds similarly, along with the amygdala and some areas of the cortex. And the same regions react in compulsive gamblers who are shown images of slot machines, suggesting that the VTA-accumbens pathway has a similarly critical role even in nondrug addictions.

DOPAMINE, PLEASE

How is it possible that diverse addictive substances—which have no common structural features and exert a variety of effects on the body—all elicit similar responses in the brain's reward circuitry? How can cocaine, a stimulant that causes the heart to race, and heroin, a pain-relieving sedative, be so opposite in some ways and yet alike in targeting the reward system? The answer is that all drugs of abuse, in addition to any other effects, cause the nucleus accumbens to receive a flood of dopamine and sometimes also dopamine-mimicking signals.

When a nerve cell in the VTA is excited, it sends an electrical message racing along its axon—the signal-carrying "highway" that extends into the nucleus accumbens. The signal causes dopamine to be released from the axon tip into the tiny space—the synaptic cleft—that separates the axon terminal from a neuron in the nucleus accumbens. From there, the dopamine latches on to its receptor on the accumbens neuron and transmits its signal into the cell. To later shut down the signal, the VTA neuron removes the dopamine from the synaptic cleft and repackages it to be used again as needed.

Cocaine and other stimulants temporarily disable the transporter protein that returns the neurotransmitter to the VTA neuron terminals, thereby leaving excess dopamine to act on the nucleus accumbens. Heroin and other opiates, on the other hand, bind to neurons in the VTA that normally shut down the dopamine-producing VTA neurons. The opiates release this cellular clamp, thus freeing the dopamine-secreting cells to pour extra dopamine into the nucleus accumbens. Opiates can also generate a strong "reward" message by acting directly on the nucleus accumbens.

But drugs do more than provide the dopamine jolt that induces euphoria and mediates the initial reward and reinforcement. Over time and with repeated exposure, they initiate the gradual adaptations in the reward circuitry that give rise to addiction.

AN ADDICTION IS BORN

The early stages of addiction are characterized by tolerance and dependence. After a drug binge, an addict needs more of the substance to get the same effect on mood or concentration and so on. This tolerance then provokes an escalation of drug use that engenders dependence—a need that manifests itself as painful emotional and, at times, physical reactions if access to a drug is cut off. Both tolerance and dependence occur because frequent drug use can suppress parts of the brain's reward circuit.

At the heart of this cruel suppression lies a molecule known as CREB (cAMP response element-binding protein). CREB is a transcription factor, a protein that regulates the expression, or activity, of genes and thus the overall behavior of nerve cells. When drugs of abuse are administered, dopamine concentrations

INSIGHTS FROM IMAGING

Spots of color in brain scans of cocaine addicts (below) confirm animal studies indicating that drug intake can induce profound immediate activity changes in many brain regions, including those shown; brightest spots show the most significant change. While being scanned, the subjects rated their feelings of rush and craving on a scale of zero to three—revealing that the VTA and the sublenticular extended amygdala are important to the cocaine-induced rush and that the amygdala and the nucleus accumbens influence both the rush and the craving for more drug, which becomes stronger as the euphoria wears off (graph).

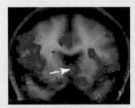
Nucleus accumbens

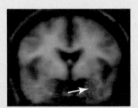
Amygdala

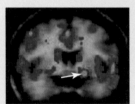
Sublenticular extended amygdala

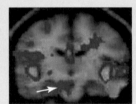
Ventral tegmental area

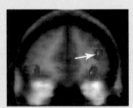
Prefrontal cortex

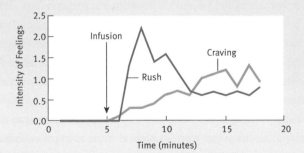

Micrographs of nucleus accumbens neurons in animals exposed to non-addictive drugs display dendritic branches with normal numbers of signal-receiving projections called spines (left and center). But those who become addicted to cocaine sprout additional spines on the branches, which consequently look bushier (right). Presumably, such remodeling makes neurons more sensitive to signals from the VTA and elsewhere and thus contributes to drug sensitivity. Recent findings suggest that delta FosB plays a part in spine growth.

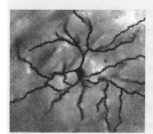

in the nucleus accumbens rise, inducing dopamine-responsive cells to increase production of a small signaling molecule, cyclic adenosine monophosphate (cAMP), which in turn activates CREB. After CREB is switched on, it binds to a specific set of genes, triggering production of the proteins those genes encode.

Chronic drug use causes sustained activation of CREB, which enhances expression of its target genes, some of which code for proteins that then dampen the reward circuitry. For example, CREB controls the production of dynorphin, a natural molecule with opiumlike effects. Dynorphin is synthesized by a subset of neurons in the nucleus accumbens that loop back and inhibit neurons in the VTA. Induction of dynorphin by CREB thereby stifles the brain's reward circuitry, inducing tolerance by making the same old dose of drug less rewarding. The increase in dynorphin also contributes to dependence, as its inhibition of the reward pathway leaves the individual, in the drug's absence, depressed and unable to take pleasure in previously enjoyable activities.

But CREB is only a piece of the story. This transcription factor is switched off within days after drug use stops. So CREB cannot account for the longer-lasting grip that abused substances have on the brain—for the brain alterations that cause addicts to return to a substance even after years or decades of abstinence.

Such relapse is driven to a large extent by sensitization, a phenomenon whereby the effects of a drug are augmented.

Although it might sound counterintuitive, the same drug can evoke both tolerance and sensitization. Shortly after a hit, CREB activity is high and tolerance rules: for several days, the user would need increasing amounts of drug to goose the reward circuit. But if the addict abstains, CREB activity declines. At that point, tolerance wanes and sensitization sets in, kicking off the intense craving that underlies the compulsive drug-seeking behavior of addiction. A mere taste or a memory can draw the addict back. This relentless yearning persists even after long periods of abstention. To understand the roots of sensitization, we have to look for molecular changes that last longer than a few days. One candidate culprit is another transcription factor: delta FosB.

ROAD TO RELAPSE

Delta FosB appears to function very differently in addiction than CREB does. Studies of mice and rats indicate that in response to chronic drug abuse, delta FosB concentrations rise gradually and progressively in the nucleus accumbens and other brain regions. Moreover, because the protein is extraordinarily stable, it remains active in these nerve cells for weeks to months after drug administration, a persistence that would enable it to maintain changes in gene expression long after drug taking ceased.

Studies of mutant mice that produce excessive amounts of delta FosB in the nucleus accumbens show that prolonged induction of this molecule causes animals to become hypersensitive to drugs. These mice were highly prone to relapse after the drugs were withdrawn and later made available—a finding implying that delta FosB concentrations could well contribute to long-term increases in sensitivity in the reward pathways of humans. Interestingly, delta FosB is also produced in the nucleus accumbens in mice in response to repetitious nondrug rewards, such as excessive wheel running and sugar consumption. Hence, it might have a more general role in the development of compulsive behavior toward a wide range of rewarding stimuli.

Recent evidence hints at a mechanism for how sensitization could persist even after delta FosB concentrations return to normal. Chronic exposure to cocaine and other drugs of abuse is known to induce the signal-receiving branches of nucleus accumbens neurons to sprout additional buds, termed dendritic spines, that bolster the cells' connections to other neurons. In rodents, this sprouting can continue for some months after drug taking ceases. This discovery suggests that delta FosB may be responsible for the added spines. Highly speculative extrapolation from these results raises the possibility that the extra connections generated by delta FosB activity amplify signaling between the linked cells for years and that such heightened signaling might cause the brain to overreact to drug-related cues. The dendritic changes may, in the end, be the key adaptation that accounts for the intransigence of addiction.

LEARNING ADDICTION

Thus far we have focused on drug-induced changes that relate to dopamine in the brain's reward system. Recall, however, that other brain regions—namely, the amygdala, hippocampus and frontal cortex—are involved in addiction and communicate back and forth with the VTA and the nucleus accumbens. All those regions talk to the reward pathway by releasing the neurotransmitter glutamate. When drugs of abuse increase dopamine release from the VTA into the nucleus accumbens, they also alter the responsiveness of the VTA and nucleus accumbens to glutamate for days. Animal experiments indicate that changes in sensitivity to glutamate in the reward pathway enhance both the release of dopamine from the VTA and responsiveness to dopamine in the nucleus accumbens, thereby promoting CREB and delta FosB activity and the unhappy effects of these molecules. Furthermore, it seems that this altered glutamate sensitivity strengthens the neuronal pathways that link memories of drug-taking experiences with high reward, thereby feeding the desire to seek the drug.

The mechanism by which drugs alter sensitivity to glutamate in neurons of the reward pathway is not yet known with certainty, but a working hypothesis can be formulated based on how glutamate affects neurons in the hippocampus. There certain types of short-term stimuli can enhance a cell's response to glutamate over many hours. The phenomenon, dubbed long-term potentiation, helps memories to form and appears to be mediated by the shuttling of certain glutamate-binding receptor proteins from intracellular stores, where they are not functional, to the nerve cell membrane, where they can respond to glutamate released into a synapse. Drugs of abuse influence the shuttling of glutamate receptors in the reward pathway. Some findings suggest that they can also influence the synthesis of certain glutamate receptors.

Taken together, all the drug-induced changes in the reward circuit that we have discussed ultimately promote tolerance, dependence, craving, relapse and the complicated behaviors that accompany addiction. Many details remain mysterious, but we can say some things with assurance. During prolonged drug

DIFFERENT DRUGS, SAME ULTIMATE EFFECT

Drugs of abuse hit various targets in the brain, but all directly or indirectly enhance the amount of dopamine signaling in the nucleus accumbens, thereby promoting addiction. Knowledge of the targets raises ideas for therapy.

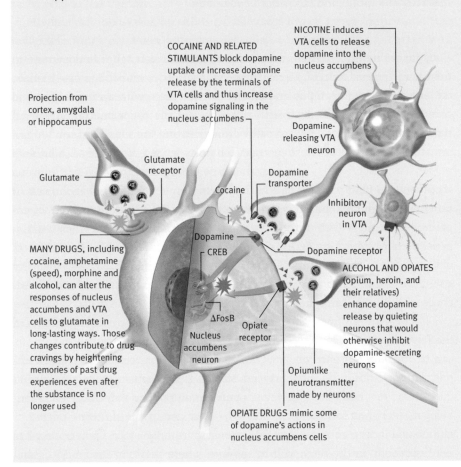

NICOTINE induces VTA cells to release dopamine into the nucleus accumbens

COCAINE AND RELATED STIMULANTS block dopamine uptake or increase dopamine release by the terminals of VTA cells and thus increase dopamine signaling in the nucleus accumbens

Projection from cortex, amygdala or hippocampus

Dopamine-releasing VTA neuron

Glutamate receptor

Glutamate

Dopamine transporter

Cocaine

Inhibitory neuron in VTA

Dopamine

CREB

Dopamine receptor

MANY DRUGS, including cocaine, amphetamine (speed), morphine and alcohol, can alter the responses of nucleus accumbens and VTA cells to glutamate in long-lasting ways. Those changes contribute to drug cravings by heightening memories of past drug experiences even after the substance is no longer used

ΔFosB

Nucleus accumbens neuron

Opiate receptor

ALCOHOL AND OPIATES (opium, heroin, and their relatives) enhance dopamine release by quieting neurons that would otherwise inhibit dopamine-secreting neurons

Opiumlike neurotransmitter made by neurons

OPIATE DRUGS mimic some of dopamine's actions in nucleus accumbens cells

use, and shortly after use ceases, changes in the concentrations of cyclic AMP and the activity of CREB in neurons in the reward pathway predominate. These alterations cause tolerance and dependence, reducing sensitivity to the drug and rendering the addict depressed and lacking motivation. With more prolonged abstention, changes in delta FosB activity and glutamate signaling predominate. These actions seem to be the ones that draw an addict back for more—by increasing sensitivity to the drug's effects if it is used again after a lapse and by

TREATMENT POSSIBILITIES

Hypothetical anticocaine agent might reduce dopamine signaling in the nucleus accumbens by interfering with cocaine's ability to block dopamine uptake by VTA neuron terminals.

Hypothetical broad-spectrum agent would mute dopamine's effects by preventing CREB or ΔFosB from accumulating or from activating the target genes of these molecules.

Hypothetical broad-spectrum agent might interfere with the unhelpful changes in glutamate signaling that occur in nucleus accumbens cells with chronic drug use.

Opiate antagonists (such as naltrexone), already on the market, block opiate receptors. They are used against alcoholism and cigarette smoking because alcohol and nicotine trigger release of the brain's own opiumlike molecules.

eliciting powerful responses to memories of past highs and to cues that bring those memories to mind.

The revisions in CREB, delta FosB and glutamate signaling are central to addiction, but they certainly are not the whole story. As research progresses, neuroscientists will surely uncover other important molecular and cellular adaptations in the reward circuit and in related brain areas that will illuminate the true nature of addiction.

A COMMON CURE?

Beyond improving understanding of the biological basis of drug addiction, the discovery of these molecular alterations provides novel targets for the biochemical treatment of this disorder. And the need for fresh therapies is enormous. In addition to addiction's obvious physical and psychological damage, the condition is a leading cause of medical illness. Alcoholics are prone to cirrhosis of the liver, smokers are susceptible to lung cancer, and heroin addicts spread HIV when they share needles. Addiction's toll on health and productivity in the U.S. has been estimated at more than $300 billion a year, making it one of the most serious problems facing society. If the definition of addiction is broadened to encompass other forms of compulsive pathological behavior, such as overeating and gambling, the costs are far higher. Therapies that could correct aberrant, addictive reactions to rewarding stimuli—whether cocaine or cheesecake or the thrill of winning at blackjack—would provide an enormous benefit to society.

Today's treatments fail to cure most addicts. Some medications prevent the drug from getting to its target. These measures leave users with an "addicted brain" and intense drug craving. Other medical interventions mimic a drug's effects and thereby dampen craving long enough for an addict to kick the habit. These chemical substitutes, however, may merely replace one habit with another. And although nonmedical, rehabilitative treatments—such as the popular 12-step programs—help many people grapple with their addictions, participants still relapse at a high rate.

Armed with insight into the biology of addiction, researchers may one day be able to design medicines that counter or compensate for the long-term effects of drugs of abuse on reward regions in the brain. Compounds that interact specifically with the receptors that bind to glutamate or dopamine in the nucleus accumbens, or chemicals that prevent CREB or delta FosB from acting on their target genes in that area, could potentially loosen a drug's grip on an addict.

Furthermore, we need to learn to recognize individuals who are most prone to addiction. Although psychological, social and environmental factors certainly are important, studies in susceptible families suggest that in humans about 50 percent of the risk for drug addiction is genetic. The particular genes involved have not yet been identified, but if susceptible individuals could be recognized early on, interventions could be targeted to this vulnerable population.

Because emotional and social factors operate in addiction, we cannot expect medications to fully treat the syndrome of addiction. But we can hope that future therapies will dampen the intense biological forces—the dependence, the cravings—that drive addiction and will thereby make psychosocial interventions more effective in helping to rebuild an addict's body and mind.

MORE TO EXPLORE

Goldstein, A. 2001. *Addiction: From biology to drug policy.* 2nd ed. Oxford University Press.

National Institute on Drug Abuse Information on Common Drugs of Abuse. www.nida.nih.gov/DrugPages

Nestler, Eric J. 2001. Molecular basis of long-term plasticity underlying addiction. *Nature Reviews Neuroscience* 2 (2): 119–128.

Robinson, Terry E., and Kent C. Berridge. 2001. Incentive-sensitization and addiction. *Addiction* 96 (1): 103–114.

ERIC J. NESTLER and ROBERT C. MALENKA study the molecular basis of drug addiction. Nestler, professor in and chair of the Department of Psychiatry at the University of Texas Southwestern Medical Center at Dallas, was elected to the Institute of Medicine in 1998. Malenka, professor of psychiatry and behavioral sciences at the Stanford University School of Medicine, joined the faculty there after serving as director of the Center for the Neurobiology of Addiction at the University of California, San Francisco. With Steven E. Hyman, now at Harvard University, Nestler and Malenka wrote the textbook *Molecular Basis of Neuropharmacology* (McGraw-Hill, 2001).

Originally published in *Scientific American*, Vol. 290, No. 3, March 2004.

14

Decoding Schizophrenia

A fuller understanding of signaling in the brains of people with this disorder
offers new hope for improved therapy

DANIEL C. JAVITT AND JOSEPH T. COYLE

Today the word "schizophrenia" brings to mind such names as John Nash and Andrea Yates. Nash, the subject of the Oscar-winning film *A Beautiful Mind,* emerged as a mathematical prodigy and eventually won a Nobel Prize for his early work, but he became so profoundly disturbed by the brain disorder in young adulthood that he lost his academic career and floundered for years before recovering. Yates, a mother of five who suffers from both depression and schizophrenia, infamously drowned her young children in a bathtub to "save them from the devil" and is now in prison.

The experiences of Nash and Yates are typical in some ways but atypical in others. Of the roughly 1 percent of the world's population stricken with schizophrenia, most remain largely disabled throughout adulthood. Rather than being geniuses like Nash, many show below-average intelligence even before they become symptomatic and then undergo a further decline in IQ when the illness sets in, typically during young adulthood. Unfortunately, only a minority ever achieve gainful employment. In contrast to Yates, fewer than half marry or raise families. Some 15 percent reside for long periods in state or county mental health facilities, and another 15 percent end up incarcerated for petty crimes and vagrancy. Roughly 60 percent live in poverty, with one in 20 ending up homeless. Because of poor social support, more individuals with schizophrenia become victims than perpetrators of violent crime.

Medications exist but are problematic. The major options today, called antipsychotics, stop all symptoms in only about 20 percent of patients. (Those lucky enough to respond in this way tend to function well as long as they continue

OVERVIEW/SCHIZOPHRENIA

- Scientists have long viewed schizophrenia as arising out of a disturbance in a particular brain system—one in which brain cells communicate using a signaling chemical, or neurotransmitter, called dopamine.
- Yet new research is shifting emphasis from dopamine to another neurotransmitter, glutamate. Impaired glutamate signaling appears to be a major contributor to the disorder.
- Drugs are now in development to treat the illness based on this revised understanding of schizophrenia's underlying causes.

treatment; too many, however, abandon their medicines over time, usually because of side effects, a desire to be "normal" or a loss of access to mental health care). Two thirds gain some relief from antipsychotics yet remain symptomatic throughout life, and the remainder show no significant response.

An inadequate arsenal of medications is only one of the obstacles to treating this tragic disorder effectively. Another is the theories guiding drug therapy. Brain cells (neurons) communicate by releasing chemicals called neurotransmitters that either excite or inhibit other neurons. For decades, theories of schizophrenia have focused on a single neurotransmitter: dopamine. In the past few years, though, it has become clear that a disturbance in dopamine levels is just a part of the story and that, for many, the main abnormalities lie elsewhere. In particular, suspicion has fallen on deficiencies in the neurotransmitter glutamate. Scientists now realize that schizophrenia affects virtually all parts of the brain and that, unlike dopamine, which plays an important role only in isolated regions, glutamate is critical virtually everywhere. As a result, investigators are searching for treatments that can reverse the underlying glutamate deficit.

MULTIPLE SYMPTOMS

To develop better treatments, investigators need to understand how schizophrenia arises—which means they need to account for all its myriad symptoms. Most of these fall into categories termed "positive," "negative" and "cognitive." Positive symptoms generally imply occurrences beyond normal experience; negative symptoms generally connote diminished experience. Cognitive, or "disorganized," symptoms refer to difficulty maintaining a logical, coherent flow of conversation, maintaining attention, and thinking on an abstract level.

The public is most familiar with the positive symptoms, particularly agitation, paranoid delusions (in which people feel conspired against) and hallucinations, commonly in the form of spoken voices. Command hallucinations, where voices tell people to hurt themselves or others, are an especially ominous sign: they can be difficult to resist and may precipitate violent actions.

The negative and cognitive symptoms are less dramatic but more pernicious. These can include a cluster called the four A's: autism (loss of interest in other people or the surroundings), ambivalence (emotional withdrawal), blunted affect (manifested by a bland and unchanging facial expression), and the cognitive problem of loose association (in which people join thoughts without clear logic, frequently jumbling words together into a meaningless word salad). Other common symptoms include a lack of spontaneity, impoverished speech, difficulty establishing rapport and a slowing of movement. Apathy and disinterest especially can cause friction between patients and their families, who may view these attributes as signs of laziness rather than manifestations of the illness.

When individuals with schizophrenia are evaluated with pencil-and-paper tests designed to detect brain injury, they show a pattern suggestive of widespread dysfunction. Virtually all aspects of brain operation, from the most basic sensory processes to the most complex aspects of thought, are affected to some extent. Certain functions, such as the ability to form new memories either temporarily or permanently or to solve complex problems, may be particularly impaired. Patients also display difficulty solving the types of problems encountered in daily living, such as describing what friends are for or what to do if all the lights in the house go out at once. The inability to handle these common problems, more than anything else, accounts for the difficulty such individuals have in living independently. Overall, then, schizophrenia conspires to rob people of the very qualities they need to thrive in society: personality, social skills and wit.

BEYOND DOPAMINE

The emphasis on dopamine-related abnormalities as a cause of schizophrenia emerged in the 1950s, as a result of the fortuitous discovery that a class of medication called the phenothiazines was able to control the positive symptoms of the disorder. Subsequent studies demonstrated that these substances work by blocking the functioning of a specific group of chemical-sensing molecules called dopamine D2 receptors, which sit on the surface of certain nerve cells and convey dopamine's signals to the cells' interior. At the same time, research led by the recent Nobel laureate Arvid Carlsson revealed that amphetamine,

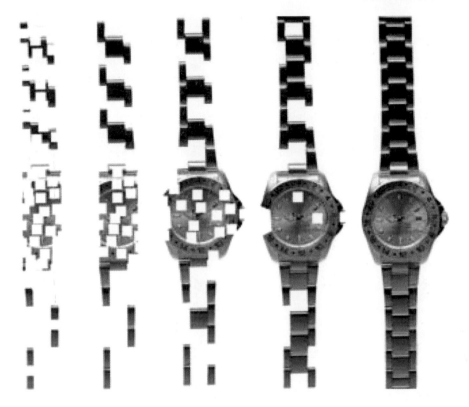

Perceiving fragments as parts of a whole can be difficult for people with schizophrenia. When normal subjects view fractured images like those above in sequence, they identify the object quickly, but schizophrenic patients often cannot make that leap swiftly.

which was known to induce hallucinations and delusions in habitual abusers, stimulated dopamine release in the brain. Together these two findings led to the "dopamine theory," which proposes that most symptoms of schizophrenia stem from excess dopamine release in important brain regions, such as the limbic system (thought to regulate emotion) and the frontal lobes (thought to regulate abstract reasoning).

Over the past 40 years, both the strengths and the limitations of the theory have become apparent. For some patients, especially those with prominent positive symptoms, the theory has proved robust, fitting symptoms and guiding treatment well. The minority of those who display only positive manifestations frequently function quite well—holding jobs, having families and

THE BRAIN IN SCHIZOPHRENIA

Many brain regions and systems operate abnormally in schizophrenia, including those highlighted below. Imbalances in the neurotransmitter dopamine were once thought to be the prime cause of schizophrenia. But new findings suggest that impoverished signaling by the more pervasive neurotransmitter glutamate—or, more specifically, by one of glutamate's key targets on neurons (the NMDA receptor)—better explains the wide range of symptoms in this disorder.

AUDITORY SYSTEM
Enables humans to hear and understand speech. In schizophrenia, overactivity of the speech area (called Wernicke's area) can create auditory hallucinations—the illusion that internally generated thoughts are real voices coming from the outside

BASAL GANGLIA
Involved in movement and emotions and in Integrating sensory information. Abnormal functioning in schizophrenia is thought to contribute to paranoia and hallucinations. (Excessive blockade of dopamine receptors in the basal ganglia by traditional antipsychotic medicines leads to motor side effects)

OCCIPITAL LOBE
Processes information about the visual world. People with schizophrenia rarely have full-blown visual hallucinations, but disturbances in this area contribute to such difficulties as interpreting complex images, recognizing motion, and reading emotions on others' faces

FRONTAL LOBE
Critical to problem solving, insight and other high-level reasoning. Perturbations in schizophrenia lead to difficulty in planning actions and organizing thoughts

HIPPOCAMPUS
Mediates learning and memory formation, intertwined functions that are impaired in schizophrenia

LIMBIC SYSTEM
Involved in emotion. Disturbances are thought to contribute to the agitation frequently seen in schizophrenia

suffering relatively little cognitive decline over time—if they stick with their medicines.

Yet for many, the hypothesis fits poorly. These are the people whose symptoms come on gradually, not dramatically, and in whom negative symptoms overshadow the positive. The sufferers grow withdrawn, often isolating themselves for years. Cognitive functioning is poor, and patients improve slowly, if at all, when treated with even the best existing medications on the market.

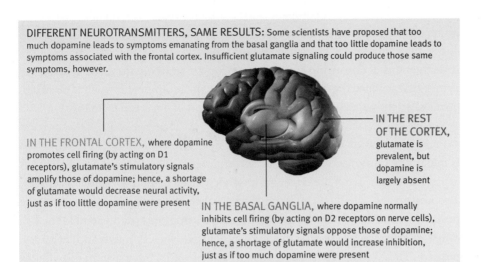

DIFFERENT NEUROTRANSMITTERS, SAME RESULTS: Some scientists have proposed that too much dopamine leads to symptoms emanating from the basal ganglia and that too little dopamine leads to symptoms associated with the frontal cortex. Insufficient glutamate signaling could produce those same symptoms, however.

IN THE FRONTAL CORTEX, where dopamine promotes cell firing (by acting on D1 receptors), glutamate's stimulatory signals amplify those of dopamine; hence, a shortage of glutamate would decrease neural activity, just as if too little dopamine were present

IN THE REST OF THE CORTEX, glutamate is prevalent, but dopamine is largely absent

IN THE BASAL GANGLIA, where dopamine normally inhibits cell firing (by acting on D2 receptors on nerve cells), glutamate's stimulatory signals oppose those of dopamine; hence, a shortage of glutamate would increase inhibition, just as if too much dopamine were present

Such observations have prompted some researchers to modify the dopamine hypothesis. One revision suggests, for example, that the negative and cognitive symptoms may stem from *reduced* dopamine levels in certain parts of the brain, such as the frontal lobes, and increased dopamine in other parts of the brain, such as the limbic system. Because dopamine receptors in the frontal lobe are primarily of the D1 (rather than D2) variety, investigators have begun to search, so far unsuccessfully, for medications that stimulate D1 receptors while inhibiting D2s.

In the late 1980s researchers began to recognize that some pharmaceuticals, such as clozapine (Clozaril), were less likely to cause stiffness and other neurologic side effects than older treatments, such as chlorpromazine (Thorazine) or haloperidol (Haldol), and were more effective in treating persistent positive and negative symptoms. Clozapine, known as an atypical antipsychotic, inhibits dopamine receptors less than the older medications and affects the activity of various other neurotransmitters more strongly. Such discoveries led to the development and wide adoption of several newer atypical antipsychotics based on the actions of clozapine (certain of which, unfortunately, now turn out to be capable of causing diabetes and other unexpected side effects). The discoveries also led to the proposal that dopamine was not the only neurotransmitter disturbed in schizophrenia; others were involved as well.

Theories focusing largely on dopamine are problematic on additional grounds. Improper dopamine balance cannot account for why one individual with schizophrenia responds almost completely to treatment, whereas someone else shows no apparent response. Nor can it explain why positive symptoms respond so much better than negative or cognitive ones do. Finally, despite decades of research, investigations of dopamine have yet to uncover a smoking gun. Neither the enzymes that produce this neurotransmitter nor the receptors to which it binds appear sufficiently altered to account for the panoply of observed symptoms.

THE ANGEL DUST CONNECTION

If dopamine cannot account well for schizophrenia, what is the missing link? A critical clue came from the effects of another abused drug: PCP (phencyclidine), also known as angel dust. In contrast to amphetamine, which mimics only the positive symptoms of the disease, PCP induces symptoms that resemble the full range of schizophrenia's manifestations: negative and cognitive and, at times, positive. These effects are seen not just in abusers of PCP but also in individuals given brief, low doses of PCP or ketamine (an anesthetic with similar effects) in controlled drug-challenge trials.

Such studies first drew parallels between the effects of PCP and the symptoms of schizophrenia in the 1960s. They showed, for example, that individuals receiving PCP exhibited the same type of disturbances in interpreting proverbs as those with schizophrenia. More recent studies with ketamine have produced even more compelling similarities. Notably, during ketamine challenge, normal individuals develop difficulty thinking abstractly, learning new information, shifting strategies or placing information in temporary storage. They show a general motor slowing and reduction in speech output just like that seen in schizophrenia. Individuals given PCP or ketamine also grow withdrawn, sometimes even mute; when they talk, they speak tangentially and concretely. PCP and ketamine rarely induce schizophrenia-like hallucinations in normal volunteers, but they exacerbate these disturbances in those who already have schizophrenia.

The ability of PCP and ketamine to induce a broad spectrum of schizophrenia-like symptoms suggests that these drugs replicate some key molecular disturbance in the brain of schizophrenic patients. At the molecular level the drugs impair the functioning of the brain signaling systems that rely on glutamate, the main excitatory neurotransmitter in the brain. More precisely, they block the action of a form of glutamate receptor known as the NMDA receptor, which plays a critical role in brain development, learning, memory and neural processing in general. This receptor also participates in regulating dopamine release, and blockade of NMDA receptors produces the same disturbances of dopamine function typically seen in schizophrenia. Thus, NMDA receptor dysfunction, by

DRUG CLASSES IN DEVELOPMENT

Unless otherwise noted, the compounds mentioned below are in the early stages of human testing. Their developers or producers are listed in parentheses.

- Stimulators of NMDA-type glutamate receptors aim to overcome the signaling deficits that apparently contribute to many schizophrenic symptoms. Examples: Glycine (Medifoods), D-serine (Glytech). As natural substances, both of them are sold, but they remain under evaluation specifically for their value in treating schizophrenia.

- Stimulators of AMPA-type glutamate receptors—also called ampakines—may improve some aspects of memory and cognition in people with schizophrenia. Example: CX516 (Cortex Pharmaceuticals)

- Modulators of another class of glutamate receptors—metabotropic receptors—can regulate glutamate release and potentially restore the balance between the activity of NMDA and AMPA receptors. Example: LY354740 (Eli Lilly)

- Inhibitors of glycine transport reduce glycine removal from synapses, which should increase signaling by NMDA receptors. Example: GlyT-1 (NPS Pharmaceuticals and Janssen Pharmaceutica)

- Stimulators of alpha 7 nicotinic receptors, the same receptors activated by the nicotine in cigarettes, indirectly stimulate the brain's NMDA receptors. Schizophrenics often smoke heavily, probably because the nicotine, acting on alpha 7 receptors, helps them to focus. Example: DMXB-A (University of Colorado Health Sciences Center)

- Stimulators of D1 dopamine receptors are being developed mainly for Parkinson's disease and have passed initial safety trials. They might also correct dopamine deficiencies in schizophrenia, but clinical trials for that purpose have not yet been performed. Example: ABT-431 (Abbott Laboratories)

itself, can explain both negative and cognitive symptoms of schizophrenia as well as the dopamine abnormalities at the root of the positive symptoms.

One example of the research implicating NMDA receptors in schizophrenia relates to the way the brain normally processes information. Beyond strengthening connections between neurons, NMDA receptors amplify neural signals,

STEEP SOCIAL COSTS

Schizophrenia, which affects about two million Americans, takes an enormous toll on society. Because it tends to arise in young adulthood and persist, it rings up a huge tally in health care bills and lost wages and ranks among the costliest illnesses in the U.S.

Treatment and strong social support enable some individuals to lead relatively productive and satisfying lives, but most are not so lucky. Fewer than one-third can hold a job, and half of those do so only because they have intensive assistance. Men (who tend to become symptomatic earlier than women) usually do not marry, and women who tie the knot frequently enter into marriages that do not last. Because individuals with schizophrenia often isolate themselves and lack jobs, they constitute a disproportionate share of the chronically homeless population.

People with this disorder also have a high likelihood of becoming substance abusers. About 60 percent of symptomatic individuals smoke cigarettes, and half abuse alcohol, marijuana or cocaine. Such activities can lead to poor compliance with treatment and can exacerbate psychotic symptoms, increasing propensities toward violence. (Abstainers, however, behave no more violently than the general population.) Homelessness and substance abuse combine to land many with schizophrenia in prisons and county jails, where they often fail to get the treatment they require.

The grim figures do not end there: roughly 10 percent of people with schizophrenia commit suicide (usually during the illness's early stages), a higher rate than results from major depression. But there is one bright note: clozapine, the atypical antipsychotic introduced in 1989, has recently been shown to reduce the risk of suicide and substance abuse. Whether newer atypical agents exert a similar effect remains to be determined, however.

—D.C.J. and J.T.C.

much as transistors in old-style radios boosted weak radio signals into strong sounds. By selectively amplifying key neural signals, these receptors help the brain respond to some messages and ignore others, thereby facilitating mental focus and attention. Ordinarily, people respond more intensely to sounds presented infrequently than to those presented frequently and to sounds heard while listening than to sounds they make themselves while speaking. But people with schizophrenia do not respond this way, which implies that their brain circuits reliant on NMDA receptors are out of kilter.

If reduced NMDA receptor activity prompts schizophrenia's symptoms, what then causes this reduction? The answer remains unclear. Some reports show that people with schizophrenia have fewer NMDA receptors, although the genes that give rise to the receptors appear unaffected. If NMDA receptors are intact and present in proper amounts, perhaps the problem lies with a flaw in glutamate release or with a buildup of compounds that disrupt NMDA activity.

Some evidence supports each of these ideas. For instance, postmortem studies of schizophrenic patients reveal not only lower levels of glutamate but also higher levels of two compounds (NAAG and kynurenic acid) that impair the activity of NMDA receptors. Moreover, blood levels of the amino acid homocysteine are elevated; homocysteine, like kynurenic acid, blocks NMDA receptors in the brain. Overall, schizophrenia's pattern of onset and symptoms suggest that chemicals disrupting NMDA receptors may accumulate in sufferers' brains, although the research verdict is not yet in. Entirely different mechanisms may end up explaining why NMDA receptor transmission becomes attenuated.

NEW TREATMENT POSSIBILITIES

Regardless of what causes NMDA signaling to go awry in schizophrenia, the new understanding—and preliminary studies in patients—offers hope that drug therapy can correct the problem. Support for this idea comes from studies showing that clozapine, one of the most effective medications for schizophrenia identified to date, can reverse the behavioral effects of PCP in animals, something that older antipsychotics cannot do. Further, short-term trials with agents known to stimulate NMDA receptors have produced encouraging results. Beyond adding support to the glutamate hypothesis, these results have enabled long-term clinical trials to begin. If proved effective in large-scale tests, agents that activate NMDA receptors will become the first entirely new class of medicines developed specifically to target the negative and cognitive symptoms of the disorder.

The two of us have conducted some of those studies. When we and our colleagues administered the amino acids glycine and D-serine to patients with their standard medications, the subjects showed a 30 to 40 percent decline in cognitive and negative symptoms and some improvement in positive symptoms. Delivery of a medication, D-cycloserine, that is primarily used for treating tuberculosis but happens to cross-react with the NMDA receptor, produced similar results. Based on such findings, the National Institute of Mental Health has organized multicenter clinical trials at four hospitals to determine the effectiveness of D-cycloserine and glycine as therapies for schizophrenia; results should be available this year. Trials of D-serine, which is not yet approved for use in the U.S., are ongoing elsewhere with encouraging preliminary results as well. These agents have also been helpful when taken with the newest generation of atypical antipsychotics, which raises the hope that therapy can be developed to control all three major classes of symptoms at once.

It may be that none of the agents tested to date has the properties needed for commercialization; for instance, the doses required may be too high. We and others are therefore exploring alternative avenues. Molecules that slow glycine's removal from brain synapses—known as glycine transport inhibitors—might enable glycine to stick around longer than usual, thereby increasing stimulation of NMDA receptors. Agents that directly activate "AMPA-type" glutamate receptors, which work in concert with NMDA receptors, are also under active investigation. And agents that prevent the breakdown of glycine or D-serine in the brain have been proposed.

MANY AVENUES OF ATTACK

Scientists interested in easing schizophrenia are also looking beyond signaling systems in the brain to other factors that might contribute to, or protect against, the disorder. For example, investigators have applied so-called gene chips to study brain tissue from people who have died, simultaneously comparing the activity of tens of thousands of genes in individuals with and without schizophrenia. So far they have determined that many genes important to signal transmission across synapses are less active in those with schizophrenia—but exactly what this information says about how the disorder develops or how to treat it is unclear.

Genetic studies in schizophrenia have nonetheless yielded intriguing findings recently. The contribution of heredity to schizophrenia has long been controversial. If the illness were dictated solely by genetic inheritance, the identical twin of a schizophrenic person would always be schizophrenic as well, because the two have the same genetic makeup. In reality, however, when one twin has schizophrenia, the identical twin has about a 50 percent chance of also being afflicted. Moreover, only about 10 percent of first-degree family members (parents, children or siblings) share the illness even though they have on average 50 percent of genes in common with the affected individual. This disparity suggests that genetic inheritance can strongly predispose people to schizophrenia but that environmental factors can nudge susceptible individuals into illness or perhaps shield them from it. Prenatal infections, malnutrition, birth complications and brain injuries are all among the influences suspected of promoting the disorder in genetically predisposed individuals.

Over the past few years, several genes have been identified that appear to increase susceptibility to schizophrenia. Interestingly, one of these genes codes for an enzyme (catechol-O-methyltransferase) involved in the metabolism of dopamine, particularly in the prefrontal cortex. Genes coding for proteins called dysbindin and neuregulin seem to affect the number of NMDA receptors in the brain. The gene for an enzyme involved in the breakdown of D-serine (D-amino acid oxidase) may exist in multiple forms, with the most active form producing an approximately fivefold increase in risk for schizophrenia. Other genes may give rise to traits associated with schizophrenia but not the disease itself. Because each gene involved in schizophrenia produces only a small increase in risk, genetic studies must include large numbers of subjects to detect an effect, and they often generate conflicting results. On the other hand, the existence of multiple genes predisposing for schizophrenia may help explain the variability of symptoms across individuals, with some people perhaps showing the greatest effect in dopamine pathways and others evincing significant involvement of other neurotransmitter pathways.

Finally, scientists are looking for clues by imaging living brains and by comparing brains of people who have died. In general, individuals with schizophrenia have smaller brains than unaffected individuals of similar age and sex. Whereas the deficits were once thought to be restricted to areas such as the brain's frontal lobe, more recent studies have revealed similar abnormalities in many brain regions: those with schizophrenia have abnormal levels of brain response while performing tasks that activate not only the frontal lobes but also other areas of the brain, such as those that control auditory and visual processing. Perhaps the most important finding to come out of recent research is that no one area of the brain is "responsible" for schizophrenia. Just as normal behavior requires the concerted action of the entire brain, the disruption of function in schizophrenia must be seen as a breakdown in the sometimes subtle interactions both within and between different brain regions.

Because schizophrenia's symptoms vary so greatly, many investigators believe that multiple factors probably cause the syndrome. What physicians diagnose as schizophrenia today may prove to be a cluster of different illnesses, with similar and overlapping symptoms. Nevertheless, as researchers more accurately discern the syndrome's neurological bases, they should become increasingly skilled at developing treatments that adjust brain signaling in the specific ways needed by each individual.

MORE TO EXPLORE

Goff, D. C., and J. T. Coyle. 2001. The emerging role of glutamate in the pathophysiology and treatment of schizophrenia. *American Journal of Psychiatry* 158 (9): 1367–1377.

Heresco-Levy, U., D. C. Javitt, M. Ermilov, C. Mordel, G. Silipo, and M. Lichtenstein. 1999. Efficacy of high-dose glycine in the treatment of enduring negative symptoms of schizophrenia. *Archives of General Psychiatry* 56 (1): 29–36.

Javitt, D. C., S. R. Zukin. 1991. Recent advances in the phencyclidine model of schizophrenia. *American Journal of Psychiatry* 148 (10): 1301–1308.

McCann, Tim, director. 2001. *Revolution #9*. Wellspring Media. VHS and DVD release, 2003.

Nasar, Sylvia. 2001. *A beautiful mind: The life of mathematical genius and Nobel laureate John Nash*. Touchstone Books.

DANIEL C. JAVITT and JOSEPH T. COYLE have studied schizophrenia for many years. Javitt is director of the Program in Cognitive Neuroscience and Schizophrenia at the Nathan Kline Institute for Psychiatric Research in Orangeburg, N.Y., and professor of psychiatry at the New York University School of Medicine. His paper demonstrating that the glutamate-blocking drug PCP reproduces the symptoms of schizophrenia was the second-most-cited schizophrenia publication of the 1990s. Coyle is Eben S. Draper Professor of Psychiatry and Neuroscience at Harvard Medical School and also editor in chief of the *Archives of General Psychiatry*. Both authors have won numerous awards for their research. Javitt and Coyle hold independent patents for use of NMDA modulators in the treatment of schizophrenia, and Javitt has significant financial interests in Medifoods and Glytech, companies attempting to develop glycine and D-serine as treatments for schizophrenia.

Originally published in *Scientific American*, Vol. 290, No. 1, January 2004.

15

Turning Off Depression

Helen Mayberg may have found the switch that lifts depression—and shined a light on the real link between thought and emotion

DAVID DOBBS

Given her background and curiosity, Helen Mayberg seems to have been destined from girlhood to do what she is doing now—even though her current work was inconceivable then. Her father practiced family medicine in Los Angeles County. Her uncle used x-rays and nuclear medicine machines to research biochemistry. Today Mayberg peers into brains to examine mood networks—and with one startling experiment has transformed the treatment of depression. At the same time, by combining her father's bedside dedication with her uncle's technical sophistication, she is changing the leading theories of how thought and mood interact.

Like many researchers, Mayberg began her career hoping to advance her discipline. She expected to do so in the usual way, by slowly accruing results that would eventually alter the landscape. Now based in Atlanta at Emory University as a professor of psychiatry and neurology, she has indeed achieved such an effect. But last year she also created a big peak all at once, when she and two collaborators described how they cured eight of 12 spectacularly depressed people—individuals virtually catatonic with depression despite years of talk therapy, drugs, even shock therapy. They did so by inserting pacemaker-like electrodes into a spot deep in the cortex known as area 25. A decade earlier Mayberg had identified area 25 as a key conduit of neural traffic between the "thinking" frontal cortex and the central limbic region that gives rise to emotion, which appeared earlier in our evolutionary development. She subsequently found that area 25 runs hot in depressed and sad people—"like a gate left open," as she puts it—allowing negative emotions to overwhelm thinking and mood. Inserting the electrodes closed this gate and rapidly alleviated the depression of two thirds of the trial's patients. The study won her instant renown. "Mayberg is beginning to do for depression what we did 25 years ago for cancer," says Thomas Insel, director of the National Institute of Mental Health. "It's early yet. But we can safely say that Mayberg's work shows us whole new avenues into understanding and treating depression."

Mayberg's success stems from a certain irony. She thinks she is probably the only board-certified neurologist whose main title is professor of psychiatry, which she says is "sort of strange" considering that she rejected psychiatry, her original choice of study, as too nebulous a discipline. "I didn't like the tool kit," she explains. Even though she says she is "all about the wiring diagram" of the brain, she has produced one of the most significant findings in years about depression, psychiatry's most common and elusive patient problem. And her discovery could redefine our understanding of the relation between reasoned thought and unreasoned emotion.

THE GRAIL: AREA 25

Sit down for dinner with Mayberg, as I did at Mayhene's in Washington, D.C., where she had come for a conference, and you are treated not just to a good meal but to intellectual excitement. Lively, with big eyes and a ready smile, Mayberg exudes the enthusiasm of a freshly inspired grad student combined with a 50-year-old veteran's appreciation of history.

"I was always a tinkerer," she recounts. "Summers I used to spend hours in my uncle's lab [at the University of California] at Berkeley. He did early work mapping out thyroxine dynamics in the brain. We'd talk mapping, which I've always found fascinating, and he'd give me little lab tasks to do. I loved the lab—the logic of it, the gadgets and Geiger counters. Measuring things to solve puzzles."

She entered medical school at the University of California, Los Angeles, figuring she would be a psychiatrist. Yet when she advanced to her psychiatry rotations in the late 1970s she found few gadgets and little quantitative measurement. "There were no CT scans available then," she recalls, "much less PET imaging or fMRIs. And most psychiatrists didn't fully accept the biology underlying psychiatric disorders." For instance, the profession viewed schizophrenia—which today is seen to rise from genetic and neural underpinnings—as primarily a reaction to maternal neglect or abuse.

In 1980 Mayberg did a senior-year clerkship with neurologist Norman Geschwind at Harvard University's Beth Israel Hospital. Geschwind had spent four decades pushing the notion that the brain works as a system of coordinated functions that arise from different regions, rather than as a single unit. Dysfunction results from breakdowns in the coordination between regions.

Geschwind's vision, buttressed by his research and brilliant readings of earlier cases from neurological literature, led the profession's move from the view of a monolithic brain, which dominated the first half of the 20th century. When Mayberg began studying with Geschwind, the emerging network model was being confirmed by an explosion of discoveries about how hormones and neurotransmitters carry messages between various brain areas. Mayberg, watching Geschwind apply these models to patients on Beth Israel's neurology wards, found a far more appealing theory of mental function than psychiatry offered.

After graduating, she took up a neurology residency at Columbia University, where she investigated depression in stroke patients. She hoped to localize the neural networks involved. But the stroke patients' lesions varied so much in location and severity that she could not find consistent patterns.

Still, the project honed her interest, and when she finished the residency and moved to a postdoctoral program at Johns Hopkins University, she began studying depression in Parkinson's patients. Parkinson's offered more promise for isolating neural networks, because it results from damage to a well-defined, deep-brain structure crucial to movement, the globus pallidus. At the time, Johns Hopkins led the world in neurotransmitter research, breaking new ground almost monthly on dopamine and serotonin function, so Mayberg naturally started by trying to find anomalies in the patients' neurochemistry. But focusing on chemicals suited her little better than psychiatry did.

"With psychiatry," she explains, "the resolution was the whole brain. That was too low resolution for me. I discovered that the chemistry"—neurotransmitter action at the cellular level—"was too fine a resolution. I wanted to see how the parts worked together."

So Mayberg, applying her uncle's discipline of nuclear medicine, developed a new project in the early 1990s. She and some collaborators scanned 60 Parkinson's patients, some depressed and some not, with positron-emission tomography (PET). They were looking for differences in activity in the frontal and paralimbic regions—the "thinking" frontal cortex behind the forehead and the "older," more interior paralimbic cortex surrounding the limbic centers for emotion, memory and learning. They found that the depressed patients showed far less activity in both cortex regions. Over the next few years Mayberg performed similar studies comparing depressed and nondepressed patients who experienced strokes or who had Huntington's, epilepsy or Alzheimer's. The depressed patients in every study had the same reduced frontal and paralimbic activity.

Mayberg also found something else: the depressed people had one particular segment of evolutionarily older cortex, just over the roof of the mouth, that was especially busy. It was the region called area 25. Another researcher working independently—Wayne Drevets of Washington University (now at the National Institute of Mental Health)—also noticed this hyperactivity. The notion seemed odd; in depression, characterized by underactivity in the brain, one localized network was overactive. Area 25 proved to have strong connections between the limbic system's emotional and memory centers and the frontal cortex's thinking centers. Exactly how area 25 modulated traffic between these districts was not clear, but the region was clearly hyperactive in cases of severe depression. Perhaps it was working overtime as it tried to temper a depressive loop set up between emotional and thinking centers. Or perhaps it actually caused the problem by kicking into overdrive and letting depressive loops take over. In any case, Mayberg says, "we were seeing area 25 as important." It suggested a pattern, something fundamental about depression.

In 1997 Mayberg wrote a long theoretical review paper describing the findings supporting this pattern. Few psychiatrists took notice. "Quite frankly," she says, "no one was particularly interested. I was asking them to look at a lot of brain regions and think of depression in a new way. People weren't ready for it. So I got put in a box."

Because most of her studies had been on people suffering some other neurological problem, such as Parkinson's or epilepsy, her colleagues branded the patients as having "secondary depression" rather than ordinary "primary depression." Their symptoms were an inevitable—and essentially unimportant—side effect of the main condition. "So they'd say, 'Oh, you do that neurological depression stuff,'" Mayberg recalls. "'Very nice.' And I'm saying, 'No, no, no! This is about all depression.' But it just seemed to annoy people."

STUMPED

Annoyance changed to attention at the century's turn, however, as Mayberg tested her assertions with increasingly revealing studies. She asked healthy subjects to think sad thoughts and scanned them when the tears were flowing. The images showed depressed frontal activity and a hyperactive area 25. Yet as the sadness passed, the frontal area revived and area 25 calmed. She scanned depressed patients undergoing treatment with Paxil or with placebos. In both groups, individuals who recovered showed a rise in frontal activity and a calming in area 25. It seemed that, no matter what the cause, depression dampened frontal activity and either caused or rose from hyperactivity in area 25. And for all afflicted, curing the depression reversed these effects.

Then, in early 2004, Mayberg published a study that drew wide notice, and her own results threw her for a loop. She scanned two groups of depressed patients undergoing treatment—one with Paxil and the other with cognitive behavioral therapy (CBT), which aims to cure through counseling. The Paxil patients showed the same pattern as the earlier studies had found. The CBT patients displayed a new and confounding dynamic, however: when CBT treatment worked, area 25 slowed down, as expected, but the frontal areas showed *less* activity. They went from heightened to lower activity, instead of low to high, as had occurred in every other group.

"Oh, *man*," Mayberg says. "I was stumped. For a while I had to just set it aside." Why did the CBT patients' frontal activity drop instead of rising as they got better? After discussions and contemplation, she finally realized the answer. The successful CBT patients, almost by definition, had to show this pattern. In CBT, patients learn to recognize and change thought patterns that would otherwise depress them. An active frontal area was virtually required to make CBT work. The patients who responded to CBT did so either because they were busier thinkers by nature (and therefore more amenable to CBT) or because they entered the study already trying to think their way out of their depression. The scans showing initial high levels of frontal activity, Mayberg explains, "were pictures of the tug-of-war between the depression and the patients' attempts to self-correct." When the attempt succeeded, the frontal areas could relax, and the scans showed the reduced activity.

This anomalous result held ripe suggestions about what kind of patients might best respond to CBT versus drug therapy. It also highlighted the central finding uniting all the various studies: even the CBT responders had an initially hyperactive area 25 that settled down as therapy worked and mood improved. Area 25 was overly busy in all types of depressions and was calmed by any successful therapy.

INSTANT RELIEF

Mayberg now possessed strong, replicated evidence that area 25 played a fundamental role in depression. This insight fit well with what others had discovered about the dynamics of fear, anxiety, stress and mood. Researchers such as New York University neuroscientist Joseph E. LeDoux (see "Mastery of Emotion," Dobbs, *Scientific American Mind*, February/March) and Bruce McEwen, a neuroendocrinologist at the Rockefeller University, had shown that mood disorders often develop because extreme or continuous stress, whether from a trauma or a difficult ongoing environment, kick fear and anxiety centers into long-term overdrive. The survival systems that have long served us well—a heightened neural and hormonal response to acute threat—turn corrosive when such memories and persistent thoughts trigger them continuously. The evidence for this dynamic was robust. But the crucial switches in the circuit remained elusive.

Maybe, Mayberg started to think, area 25 was such a switch, and tweaking it could trip the circuit out of alarm mode and back to normal.

At about this time, Mayberg took a professorship at the University of Toronto, where she met fellow faculty members Sidney Kennedy, a psychiatrist, and Andres Lozano, a neurosurgeon. Kennedy liked to explore neurological models of depression, and Lozano had gained notoriety modulating another neural network gone awry—the one responsible for Parkinson's. In the 1980s it became common for surgeons to treat severe Parkinson's by removing the globus pallidus. The cluster of neurons is a gateway in circuits that control movement, and its hyperactivity somehow threw the neurology of movement off balance, causing the tremors and rigidity that afflict Parkinson's patients. Removing the globus pallidus seemed to reduce these complications. Lozano, on the other hand, had become one of several neurosurgeons who treated the same problem not by removing the globus pallidus but by inserting next to it a tiny, low-voltage electrode. The technique, called deep brain stimulation, seemed to regulate the activity of the globus pallidus, restoring movement to near normal.

Might inserting such electrodes alongside area 25 calm it down? Mayberg, Lozano and Kennedy decided to try it. Beginning in 2003, the team implanted electrodes in area 25 in a dozen severely depressed patients. Lozano drilled a pair of nickel-size holes in the top of the skull, slid a pair of electrodes and slender leads to area 25, attached the leads to a small pacemaker sewn in under the collarbone, and turned it on. The pacemaker sent a continuous four-volt current to area 25.

The results were stunning. Some patients felt profound relief as soon as Lozano turned on the electrodes, and two thirds returned to essentially normal mood and function within months. They saw better, thought better, felt better. They talked of feeling like they were walking amid flowers, of "the noise" stopping, of a horrid weight lifting. Side effects were almost negligible.

"We still don't really understand why calming area 25 has such an effect," Mayberg says. "That comes next. But it's clear that it causes depression when it's hyperactive and that calming it can bring relief." Indeed, the results shattered doubts. Mayberg's body of work, and this latest experiment in particular, had shown that in the emerging circuit model of mood, one could identify and modulate key switches. The results emphatically confirmed the network model of the brain as well as a long history of thought and metaphor. Reason and passion, thought and emotion, were indeed linked in a loop rather than stacked in a hierarchy. Neither stood as the other's slave. They engaged in a conversation that, to be healthy, had to be rich and balanced.

FIGURING OUT WHY

The deep-brain-stimulation trial brought Mayberg fame. The renown she doesn't mind; the affirmation she likes. "It's nice," she says, "after years of writing papers people didn't finish reading, to have people pay attention. And as a scientist, this is what you really hope for: to feel like you've gripped the wheel of a really big ship and changed its direction, even a little bit."

Yet Mayberg hardly thinks she has solved the big questions of mood and mental health. She hopes to find new tools and new working models to track and treat the complex network that links thought and mood—the cortex and limbic regions—and sends us spiraling into depression when it malfunctions. Most immediately, this search means detailing how area 25 plays so crucial a role.

"I may spend the next 10 years trying to figure out what we did," she muses. "We really did this mostly by eye. I want to figure out how to better work this area. I'd like to better define the neural network—the actual wiring, if you will.

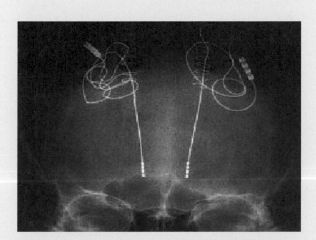

Relief: An implanted pacemaker sends current to two electrodes that reach deep into the brain, stimulating area 25 to close the open gate that feeds depression.

I'd like to map the neurochemistry more finely. I want the genetic layout. What will all that tell us about the nature of depression? Can we find more reliable differences among different types of depression? Why do some people respond to drugs and some to CBT?"

Many people would flinch at so many questions. Helen Mayberg lights up. "You know what cracks me up?" she remarks. "When people ask, 'So where are you going to look next?' I tell them, 'What do you mean, where am I going to look next? I'm going to look more closely here.'"

MORE TO EXPLORE

Goldapple, K., et al. 2004. Modulation of cortical-limbic pathways in major depression. *Archives of General Psychiatry* 61:34–41.

Mayberg, H. S., et al. 2005. Deep brain stimulation for treatment-resistant depression. *Neuron* 45 (5): 651–660.

DAVID DOBBS (www.daviddobbs.net) is a frequent contributor to *Scientific American Mind* and is the author of *Reef Madness: Charles Darwin, Alexander Agassiz, and the Meaning of Coral* (Pantheon Books, 2005).
Originally published in *Scientific American Mind*, Vol. 17, No. 4, August/September 2006.

Part 3 Tomorrow's Brain

16

Treating Depression: Pills or Talk

Medication has reduced depression for decades, but newer forms
of psychotherapy are proving their worth

STEVEN D. HOLLON, MICHAEL E. THASE,
AND JOHN C. MARKOWITZ

For decades, the public and most mental health professionals have felt that
antidepressant medications are a magic bullet for depression. Beginning in the
late 1950s, antidepressants ushered in an era of safe, reliable and reasonably affordable treatment that often produced better results than the psychotherapies
of the day. As the compounds rose in popularity, many physicians came to view
psychotherapy alone as ineffective and as little more than a minor adjunct when
combined with medication.

This is no longer the case, if it was ever true. Contrary to prevailing wisdom,
recent research suggests that several focused forms of psychotherapy may be
as effective as medication, even when treating more severe depressions. Moreover, the newer psychotherapies may provide advantages beyond what antidepressants alone can achieve. Nevertheless, pharmaceutical therapy remains the
current standard of treatment, and effective new options are being added all
the time.

These trends are important to examine because depression exacts a significant toll on society as well as individuals. Depression is one of the most common psychiatric disorders and is a leading cause of disability worldwide. The
impact of mood disorders on quality of life and economic productivity matches
that of heart disease. Depression also accounts for at least half of all suicides.

The efficacy of antidepressants has been established in thousands of placebo-
controlled trials. The newer ones are safer and have fewer noxious side effects
than earlier compounds. About 50 percent of all patients will respond to any
given medication, and many of those who do not will be helped by another agent
or a combination of them.

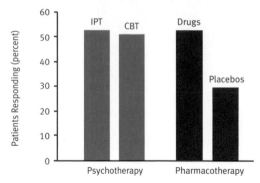

Meta-analysis indicates depression patients respond about equally well to medication or to psychotherapy (interpersonal psychotherapy—IPT—or cognitive and behavioral therapies—CBT).

Not everyone responds, however, and many who do would prefer not to have to take the pills. Quietly over the years, newer psychotherapeutic techniques have been introduced that may be just as good at alleviating acute distress in all but the most severely depressed patients. And some of the therapies provide advantages over medication alone, such as improving the quality of relationships or reducing the risk that symptoms will return after treatment is over.

This last revelation is significant because many people who recover from depression are prone to succumb again. The illness is often chronic, comparable to diabetes or hypertension, and patients treated with medication alone may have to remain on it for years, if not for life, to prevent symptoms from returning. Moreover, combining treatments—prescriptions to reduce acute symptoms quickly and psychotherapy to broaden their effects and to prevent symptoms from returning after treatment terminates—may offer the best chance for a full recovery without recurring problems.

REMISSION OR RELAPSE

Our conclusions refer mainly to the condition termed unipolar disorder. Depression comes in two basic forms: The unipolar type involves the occurrence of negative moods or loss of interest in daily activities. In the bipolar form, commonly known as manic depression, patients also experience manic states that may involve euphoria, sleeplessness, grandiosity or recklessness that can lead to everything from buying sprees to impulsive sexual adventures that later bring regret.

Bipolar disorder shows up in only 1 to 2 percent of the population and is usually treated with mood-stabilizing medication such as lithium. In contrast,

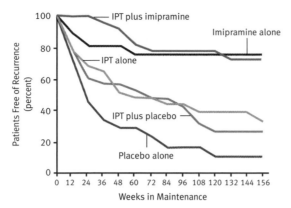

Recovered patients are least likely to suffer new depression if they continue on IPT and imipramine together (blue), rather than either treatment alone, according to one study.

about 20 percent of women and 10 percent of men suffer from unipolar depression at some time in their lives.

The treatment of unipolar depression typically progresses through three phases, determined by changes in the patient's intensity of symptoms. These are usually measured by clinical ratings such as the Hamilton Rating Scale for Depression. Seriously depressed patients in the acute phase often report feeling down much of the time. They have lost interest in formerly pleasurable activities, and they may have difficulty sleeping, changed appetite, and diminished libido. They may feel fatigued or worthless, and they may entertain recurrent thoughts of death or suicide. The goal of treatment is to relieve symptoms. "Remission" is reached when someone is fully well.

Even when in remission, however, patients may still have an elevated risk for the return of symptoms. It is common practice to encourage patients to stay on medication for at least six months following the initial remission. The return of symptoms soon after remission is called a relapse. In this sense, treating depression with drugs may be like treating an infection with antibiotics; a patient must take the medication beyond the point of first feeling better, to fully prevent the original problem from coming back. This effort to forestall relapse is called continuation treatment and typically lasts at least six to nine months beyond the point of remission.

Those who pass the point at which the treated episode is likely to return are said to have recovered. But even then, they might experience a new episode; people with a history of depression are three to five times more likely to have an episode than those with no such history. A new episode is considered a recurrence. To protect against recurrence, many patients are kept in ongoing maintenance treatment, typically medication but sometimes with psychotherapy. But once patients are off medication, having been on it does nothing to reduce subsequent risk for recurrence. Therefore, patients with a history of multiple episodes are usually advised to stay on medication indefinitely.

Although the scope of depression can vary widely, there are only a few prevailing treatments. Most of the leading antidepressants fall into three main classes: monoamine oxidase inhibitors (MAOIs), tricyclic antidepressants (TCAs) and selective serotonin reuptake inhibitors (SSRIs), such as Prozac and Paxil. Each

THE ANTIDEPRESSANTS

MAOI. Monoamine oxidase inhibitors were the first widely used antidepressants. They curtail the action of an enzyme that breaks down brain neurotransmitters. They are rarely prescribed as a first-line treatment because they require a special diet to avoid potentially dangerous, though rare, interactions with certain common foods. But they are still a medication of last resort.

TCA. Tricyclic antidepressants inhibit the reuptake of the neurotransmitters norepinephrine and serotonin. TCAs have unpleasant side effects that can include fainting, dry mouth and blurred vision; about 30 percent of patients stop taking the medication because of these problems. TCAs are also potentially lethal in an overdose. But they may still be the medication of choice for those with certain kinds of depression.

SSRI. Selective serotonin reuptake inhibitors, such as Prozac and Paxil, block the reuptake of serotonin back into presynaptic neurons. They have replaced TCAs as the primary medication because they have fewer side effects and are less likely to prove fatal in an overdose. Nevertheless, side effects such as gastrointestinal and sexual problems can be disconcerting. Indications that SSRIs may increase suicidal thoughts and actions in children and teenagers have led to mandatory warnings for these age groups in the U.S. and a ban for minors in Great Britain.

Newer medications. More doctors are trying new drugs that affect multiple neurotransmitter systems or make use of mechanisms other than blocking reuptake. Examples include bupropion, venlafaxine, nefazodone and mirtazapine. —S.H., M.T., J.M.

class has a slightly different action and different side effects and is prescribed based on a patient's history, the likelihood of certain complications and cost. Although about equally effective in a general population, some medications are more efficacious than others for specific types of depression. In general, the older MAOIs and TCAs carry greater risk of side effects than the SSRIs. But the SSRIs do not always work, especially for more severely depressed patients, and they are more expensive.

Despite the widespread use of antidepressants, their actions are not fully understood. They work in part by affecting the neurotransmitters (signaling mole-

PSYCHOTHERAPIES FOR DEPRESSION

Interpersonal psychotherapy (IPT) focuses on problems in relationships. Therapists help patients to understand life events that may have started their depression and to find ways to combat such episodes as well as reverse cycles of social withdrawal, fatigue and poor concentration. IPT emphasizes that symptoms are the result of a mood disorder and not an outgrowth of personal failure, which lifts the guilt and self-blame common in depression. **Cognitive and behavior therapies** hold that mood disorders are caused or exacerbated by learned beliefs and behaviors—which can be unlearned or modified through experience. The more cognitively based methods emphasize the role of a patient's aberrant beliefs and dysfunctional information processing, whereas the more behavioral approaches focus on how external circumstances shape patient responses. Most therapies blend cognitive and behavioral strategies and are often referred to as CBT. The goal is not to "think happy thoughts" but to become more accurate in one's self-assessments and more effective in one's behaviors. Recent variants such as mindfulness-based cognitive therapy incorporate strategies based on mediation and acceptance; others such as well-being therapy try to enhance life skills and a sense of happiness in addition to reducing distress. And still others integrate cognitive and behavioral approaches with so-called dynamic and interpersonal strategies.

More purely **behavioral therapies** akin to behavioral activation maintain that depression results from too little positive reinforcement, brought on by problems in a person's environment or a lack of social skills or a propensity to avoid challenging situations. These approaches are drawing renewed attention. —S.H., M.T., J.M.

cules in the brain) norepinephrine, serotonin and dopamine, which are involved in regulating mood, primarily by blocking the reuptake of these neurotransmitters into the neurons that secrete them. Yet this action cannot fully explain the effects, and it is quite likely that the compounds drive a subsequent cascade of biochemical events. Many people who do not respond to one antidepressant will respond to another or to a combination.

New psychotherapy methods have proved as effective as medication, although they are still not as extensively tested. The programs include interpersonal psychotherapy (IPT), which focuses on problems in relationships and helps patients lift the self-blame common in depression. Developed in the 1970s, IPT has performed well in trials but has only begun to enter clinical practice. Studies do show, however, that when IPT is paired with medication, patients receive the best of both worlds: the quick results of pharmaceutical intervention and greater breadth in improving the quality of their interpersonal lives.

Cognitive and behavioral therapies, collectively known as CBT, also compare well with medication in all but the most severely depressed patients—and they can benefit even those people if they are administered by experienced therapists. Most exciting is that CBT appears to have an enduring effect that reduces risk of relapse and perhaps recurrence. Even the most effective of the other treatments rarely have this type of long-lasting benefit. Cognitive therapy is perhaps the most well-established CBT approach. It teaches patients to examine the validity of their dysfunctional depressive beliefs and to alter how they process information about themselves. Behavioral therapy had lost favor to the cognitive approaches, but it, too, has done well in recent trials and is undergoing a revival.

WHICH WAY TO TURN

It is not possible to simply say whether medication or psychotherapy is "better" for depressed patients. But many studies have reached interesting conclusions about the approaches when they are applied across the illness's three phases: the acute symptoms at onset, the months of continuation treatment to forestall relapse, and the maintenance of health for years to come.

Among patients who take antidepressants during treatment for acute symptoms, about half show a 50 percent drop in symptom scores on rating tests over the first four to eight weeks. About one-third of those patients become fully well (remission). Not all the improvement can be attributed to pharmacology, however. In pill-placebo control experiments, placebos can achieve up to 80 percent of the success rate of active medication, probably by instilling in patients hope and the expectation for change. The placebo effect does tend to be less stable over time and smaller in magnitude in more severe or chronic depressions.

A major problem with acute-phase therapy, however, is that many stop taking their medication—primarily because of side effects—before therapists can clearly tell if the agents are working. Attrition rates from clinical trials are often 30 percent or higher for older medications such as the TCAs and around 15 percent for newer options such as the SSRIs.

The newer psychotherapies appear to do as well as medication during the acute depression phase, although the number of studies is fewer and the findings are not always consistent. One typical study found that IPT alone was about as effective as medication alone (with each better than a control condition) and that the combination was better still. In general, medication relieved symptoms more quickly, but IPT produced more improvement in social functioning and quality of relationships. The combined treatment retained the independent benefits of each.

IPT also fared well in the 1989 National Institute of Mental Health Treatment of Depression Collaborative Research Program. The TDCRP, as it is known, is perhaps the most influential study to date that compared medication and psychotherapy. In that trial, patients with major depression were randomly assigned to 16 weeks of IPT, CBT or the TCA imipramine, combined with meetings with a psychiatrist or a placebo plus meetings. Patients with less-severe depression improved equally across conditions. Among more-severely depressed patients, imipramine worked faster than IPT, but both were comparable by the end of treatment and both were superior to a placebo.

As for CBT, most of the published trials have found it to be as effective as medication in the acute phase. The most notable exception—the TDCRP—did find that cognitive therapy was less efficacious than either medication or IPT (and no better than a placebo) in the treatment of more-severely depressed patients. Because the study was large and was the first major comparison to include a pill-placebo control, its results considerably dampened enthusiasm for cognitive therapy, even though no other study had produced such a negative finding.

Today this conclusion appears to have been premature. More recent studies have found that CBT is superior to pill-placebos and is as good as an SSRI for more-severely depressed outpatients. These studies suggest that cognitive therapy's success depends greatly on the level of a therapist's training and experience with it, especially for patients with more-serious or complicated symptoms.

CONTINUING THE FIGHT

The best treatments for reducing acute distress also seem to work as well for reducing relapse when they are carried into the continuation phase. Antidepressants appear to reduce the risk for relapse by at least half. It is unclear exactly

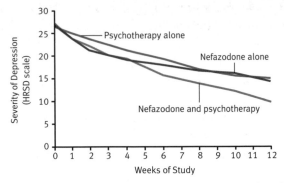

Combining medication (nefazodone) and psychotherapy (gray) reduced the intensity of symptoms most among chronically depressed patients in a 2000 study.

how long patients must keep taking medication to pass from remission into full recovery, but current convention is to go for at least six to nine months.

IPT during the continuation phase appears to prevent relapse nearly as well as medication, although studies in this regard are few. Recent investigations also suggest that if cognitive therapy is continued past the point of remission, it can reduce the risk for relapse. To date, no studies have compared continuation CBT to continuation IPT or medication.

During the maintenance phase, medication is usually recommended for high-risk patients, especially those with multiple prior episodes. Therapy can go on for years. It does protect against recurrence. Even among recovered patients, though, the risk of recurrence off medication is at least two to three times greater. Given that there is no evidence that prior medication use does anything to reduce subsequent risk for recurrence, most physicians will encourage their high-risk patients to stay on medication indefinitely.

Studies of maintenance IPT are few, but they generally support the notion that it, too, reduces risk of recurrence. It has not been as efficacious as keeping people on medication, but the handful of studies have typically cut back the frequency of IPT to monthly sessions while maintaining medication at full, acute-treatment dosages. It would be interesting to see how maintenance IPT compares when the psychotherapy sessions are also kept at "full strength."

Several studies have shown that CBT has an enduring protective benefit that extends beyond the end of treatment. Patients treated to remission with CBT were only about half as likely to relapse after treatment termination as patients treated to remission with medication, and the CBT patients were no more likely to relapse than patients who continued on the prescriptions. CBT appears to produce this enduring effect regardless of whether it is provided alone or in combination with medication during acute treatment and even if it is added only after medication has reduced acute symptoms. Further, indications are that this enduring effect may even prevent wholly new episodes (recurrence), although findings are still far from conclusive.

Given these trends, CBT may ultimately prove more cost-effective than medication. Psychotherapy usually costs at least twice as much as medication over the first several months, but if the enduring effect of CBT truly extends over time, it may prove less costly for patients to learn the skills involved and discontinue treatment than to stay on medication indefinitely. It remains unclear whether other interventions such as IPT have an enduring effect, but this possibility should certainly be explored.

Our review of the treatment literature indicates that some forms of psychotherapy can work as well as medication in alleviating acute distress. IPT may enhance the breadth of response, and CBT may enhance its stability. Combined treatment, though more costly, appears to retain the advantages of each approach. Good medical care can be hard to find, and the psychotherapies that have garnered the most empirical support are still not widely practiced. Nevertheless, some kind of treatment is almost always better than none for a person facing depression. The real tragedy is that even as alternatives expand, too few people seek help.

MORE TO EXPLORE

Frank, E., et al. 1990. Three-year outcomes for maintenance therapies in recurrent depression. *Archives of General Psychiatry* 47 (12): 1093–1099.

Hollon, Steven, Michael Thase, and John Markowitz. 2002. Treatment and prevention of depression. *Psychological Science in the Public Interest* 3 (2): 39–77.

Keller, M. B., et al. 2002. A comparison of nefazodone, the cognitive behavioral-analysis system of psychotherapy, and their combination for the treatment of chronic depression. *New England Journal of Medicine* 342 (20): 1462–1470.

STEVEN D. HOLLON, MICHAEL E. THASE, and JOHN C. MARKOWITZ study the treatment and prevention of depression. Hollon is professor of psychology at Vanderbilt University and a past president of the Association for Advancement of Behavior Therapy. Thase is professor of psychiatry at the University of Pittsburgh Medical Center and chief of adult academic psychiatry at the Western Psychiatric Institute and Clinic there. Markowitz is associate professor of psychiatry at Cornell University's Weill Medical College and a research psychiatrist at the New York State Psychiatric Institute.
Originally published in *Scientific American Mind*, Vol. 14, No.5, December 2004.

17

The Coming Merging of Mind and Machine

The accelerating pace of technological progress means that our intelligent creations will soon eclipse us—and that their creations will eventually eclipse them

RAY KURZWEIL

Sometime early in the next century, the intelligence of machines will exceed that of humans. Within several decades, machines will exhibit the full range of human intellect, emotions and skills, ranging from musical and other creative aptitudes to physical movement. They will claim to have feelings and, unlike today's virtual personalities, will be very convincing when they tell us so. By 2019 a $1,000 computer will at least match the processing power of the human brain. By 2029 the software for intelligence will have been largely mastered, and the average personal computer will be equivalent to 1,000 brains.

Once computers achieve a level of intelligence comparable to that of humans, they will necessarily soar past it. For example, if I learn French, I can't readily download that learning to you. The reason is that for us, learning involves successions of stunningly complex patterns of interconnections among brain cells (neurons) and among the concentrations of biochemicals, known as neurotransmitters, that enable impulses to travel from neuron to neuron. We have no way of quickly downloading these patterns. But quick downloading will allow our nonbiological creations to share immediately what they learn with billions of other machines. Ultimately, nonbiological entities will master not only the sum total of their own knowledge but all of ours as well.

As this happens, there will no longer be a clear distinction between human and machine. We are already putting computers—neural implants—directly into people's brains to counteract Parkinson's disease and tremors from multiple sclerosis. We have cochlear implants that restore hearing. A retinal implant is being developed in the U.S. that is intended to provide at least some visual perception for some blind individuals, basically by replacing certain visual-processing circuits of the brain. Recently scientists from Emory University implanted a chip in the brain of a paralyzed stroke victim that allows him to use his brainpower to move a cursor across a computer screen.

In the 2020s neural implants will improve our sensory experiences, memory and thinking. By 2030, instead of just phoning a friend, you will be able to meet in, say, a virtual Mozambican game preserve that will seem compellingly real. You will be able to have any type of experience—business, social, sexual—with anyone, real or simulated, regardless of physical proximity.

HOW LIFE AND TECHNOLOGY EVOLVE

To gain insight into the kinds of forecasts I have just made, it is important to recognize that technology is advancing exponentially. An exponential process starts slowly, but eventually its pace increases extremely rapidly. (A fuller documentation of my argument is contained in my new book, *The Age of Spiritual Machines*.)

The evolution of biological life and the evolution of technology have both followed the same pattern: they take a long time to get going, but advances build

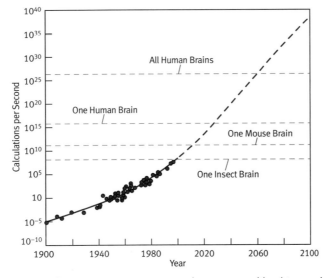

The accelerating rate of progress in computing is demonstrated by this graph, which shows the amount of computing speed that $1,000 (in constant dollars) would buy, plotted as a function of time. Computer power per unit cost is now doubling every year.

on one another and progress erupts at an increasingly furious pace. We are entering that explosive part of the technological evolution curve right now.

Consider: It took billions of years for Earth to form. It took two billion more for life to begin and almost as long for molecules to organize into the first multicellular plants and animals about 700 million years ago. The pace of evolution quickened as mammals inherited Earth some 65 million years ago. With the emergence of primates, evolutionary progress was measured in mere millions of years, leading to *Homo sapiens* perhaps 500,000 years ago.

The evolution of technology has been a continuation of the evolutionary process that gave rise to us—the technology-creating species—in the first place. It took tens of thousands of years for our ancestors to figure out that sharpening both sides of a stone created useful tools. Then, earlier in this millennium, the time required for a major paradigm shift in technology had shrunk to hundreds of years.

The pace continued to accelerate during the 19th century, during which technological progress was equal to that of the 10 centuries that came before it. Advancement in the first two decades of the 20th century matched that of the entire 19th century. Today significant technological transformations take just a few years; for example, the World Wide Web, already a ubiquitous form of communication and commerce, did not exist just nine years ago.

Computing technology is experiencing the same exponential growth. Over the past several decades, a key factor in this expansion has been described by Moore's Law. Gordon Moore, a co-founder of Intel, noted in the mid-1960s that technologists had been doubling the density of transistors on integrated circuits every 12 months. This meant computers were periodically doubling both in capacity and in speed per unit cost. In the mid-1970s Moore revised his observation of the doubling time to a more accurate estimate of about 24 months, and that trend has persisted through the 1990s.

After decades of devoted service, Moore's Law will have run its course around 2019. By that time, transistor features will be just a few atoms in width. But new computer architectures will continue the exponential growth of computing. For example, computing cubes are already being designed that will provide thousands of layers of circuits, not just one, as in today's computer chips. Other technologies that promise orders-of-magnitude increases in computing density include nanotube circuits built from carbon atoms, optical computing, crystalline computing and molecular computing.

We can readily see the march of computing by plotting the speed (in instructions per second) per $1,000 (in constant dollars) of 49 famous calculating machines spanning the 20th century. The graph is a study in exponential growth: computer speed per unit cost doubled every three years between 1910 and 1950 and every two years between 1950 and 1966 and is now doubling every year. It took 90 years to achieve the first $1,000 computer capable of executing one million instructions per second (MIPS). Now we add an additional MIPS to a $1,000 computer every day.

WHY RETURNS ACCELERATE

Why do we see exponential progress occurring in biological life, technology and computing? It is the result of a fundamental attribute of any evolutionary process, a phenomenon I call the Law of Accelerating Returns. As order exponentially increases (which reflects the essence of evolution), the time between salient events grows shorter. Advancement speeds up. The returns—the valuable products of the process—accelerate at a nonlinear rate. The escalating growth in the price performance of computing is one important example of such accelerating returns.

A frequent criticism of predictions is that they rely on an unjustified extrapolation of current trends, without considering the forces that may alter those trends. But an evolutionary process accelerates because it builds on past achievements, including improvements in its own means for further evolution. The resources it needs to continue exponential growth are its own increasing order and the chaos in the environment in which the evolutionary process takes place, which provides the options for further diversity. These two resources are essentially without limit.

The Law of Accelerating Returns shows that by 2019 a $1,000 personal computer will have the processing power of the human brain—20 million billion calculations per second. Neuroscientists came up with this figure by taking an estimation of the number of neurons in the brain, 100 billion, and multiplying it by 1,000 connections per neuron and 200 calculations per second per connection. By 2055, $1,000 worth of computing will equal the processing power of all human brains on Earth (of course, I may be off by a year or two).

PROGRAMMING INTELLIGENCE

That's the prediction for processing power, which is a necessary but not sufficient condition for achieving human-level intelligence in machines. Of greater importance is the software of intelligence.

One approach to creating this software is to painstakingly program the rules of complex processes. We are getting good at this task in certain cases; the CYC (as in "encyclopedia") system designed by Douglas B. Lenat of Cycorp has more than one million rules that describe the intricacies of human common sense, and it is being applied to Internet search engines so that they return smarter answers to our queries.

Another approach is "complexity theory" (also known as chaos theory) computing, in which self-organizing algorithms gradually learn patterns of information in a manner analogous to human learning. One such method, neural nets, is based on simplified mathematical models of mammalian neurons. Another method, called genetic (or evolutionary) algorithms, is based on allowing intelligent solutions to develop gradually in a simulated process of evolution.

Ultimately, however, we will learn to program intelligence by copying the best intelligent entity we can get our hands on: the human brain itself. We will reverse-engineer the human brain, and fortunately for us it's not even copyrighted!

The most immediate way to reach this goal is by destructive scanning: take a brain frozen just before it was about to expire and examine one very thin slice at a time to reveal every neuron, inter-neuronal connection and concentration of neurotransmitters across each gap between neurons (these gaps are called synapses). One condemned killer has already allowed his brain and body to be scanned, and all 15 billion bytes of him can be accessed on the National Library of Medicine's Web site (www.nlm.nih.gov/research/visible/visible_gallery.html). The resolution of these scans is not nearly high enough for our purposes, but the data at least enable us to start thinking about these issues.

We also have noninvasive scanning techniques, including high-resolution magnetic resonance imaging (MRI) and others. Their increasing resolution and speed will eventually enable us to resolve the connections between neurons. The rapid improvement is again a result of the Law of Accelerating Returns, because massive computation is the main element in higher-resolution imaging.

Another approach would be to send microscopic robots (or "nanobots") into the bloodstream and program them to explore every capillary, monitoring the brain's connections and neurotransmitter concentrations.

FANTASTIC VOYAGE

Although sophisticated robots that small are still several decades away at least, their utility for probing the innermost recesses of our bodies would be far-reaching. They would communicate wirelessly with one another and report their findings to other computers. The result would be a noninvasive scan of the brain taken from within.

Most of the technologies required for this scenario already exist, though not in the microscopic size required. Miniaturizing them to the tiny sizes needed, however, would reflect the essence of the Law of Accelerating Returns. For example, the translators on an integrated circuit have been shrinking by a factor of approximately 5.6 in each linear dimension every 10 years.

The capabilities of these embedded nanobots would not be limited to passive roles such as monitoring. Eventually they could be built to communicate directly with the neuronal circuits in our brains, enhancing or extending our mental capabilities. We already have electronic devices that can communicate with neurons by detecting their activity and either triggering nearby neurons to fire or suppressing them from firing. The embedded nanobots will be capable of reprogramming neural connections to provide virtual-reality experiences and to enhance our pattern recognition and other cognitive faculties.

To decode and understand the brain's information-processing methods (which, incidentally, combine both digital and analog methods), it is not necessary to see every connection, because there is a great deal of redundancy within each region. We are already applying insights from early stages of this reverse-engineering process. For example, in speech recognition, we have already decoded and copied the brain's early stages of sound processing.

Perhaps more interesting than this scanning-the-brain-to-understand-it approach would be scanning the brain for the purpose of downloading it. We would map the locations, interconnections, and contents of all the neurons, synapses and neurotransmitter concentrations. The entire organization, including the brain's memory, would then be re-created on a digital-analog computer.

To do this, we would need to understand local brain processes, and progress is already under way. Theodore W. Berger and his co-workers at the University of Southern California have built integrated circuits that precisely match the processing characteristics of substantial clusters of neurons. Carver A. Mead and his colleagues at the California Institute of Technology have built a variety of integrated circuits that emulate the digital-analog characteristics of mammalian neural circuits.

Developing complete maps of the human brain is not as daunting as it may sound. The Human Genome Project seemed impractical when it was first proposed. At the rate at which it was possible to scan genetic codes 12 years ago, it would have taken thousands of years to complete the genome. But in accordance with the Law of Accelerating Returns, the ability to sequence DNA has been

The author argues that neural implants will confer on humans an important advantage that only machines now possess: instant downloading of knowledge. Memories of events could be played back exactly as they occurred, rather than being colored by emotions. Simulations could make fantasies indistinguishable from reality.

accelerating. The latest estimates are that the entire human genome will be completed in just a few years.

By the third decade of the 21st century, we will be in a position to create complete, detailed maps of the computationally-relevant features of the human brain and to re-create these designs in advanced neural computers. We will provide a variety of bodies for our machines, too, from virtual bodies in virtual reality to bodies comprising swarms of nanobots. In fact, humanoid robots that ambulate and have lifelike facial expressions are already being developed at several laboratories in Tokyo.

WILL IT BE CONSCIOUS?

Such possibilities prompt a host of intriguing issues and questions. Suppose we scan someone's brain and reinstate the resulting "mind file" into a suitable computing medium. Will the entity that emerges from such an operation be conscious? This being would appear to others to have very much the same personality, history and memory. For some, that is enough to define consciousness. For others, such as physicist and author James Trefil, no logical reconstruction can attain human consciousness, although Trefil concedes that computers may become conscious in some new way.

At what point do we consider an entity to be conscious, to be self-aware, to have free will? How do we distinguish a process that is conscious from one that just acts *as if* it is conscious? If the entity is very convincing when it says, "I'm lonely; please keep me company," does that settle the issue?

If you ask the "person" in the machine, it will strenuously claim to be the original person. If we scan, let's say, me and reinstate that information into a neural computer, the person who emerges will think he is (and has been) me (or at least he will act that way). He will say, "I grew up in Queens, New York, went to college at MIT, stayed in the Boston area, walked into a scanner there and woke up in the machine here. Hey, this technology really works."

But wait, is this really me? For one thing, old Ray (that's me) still exists in my carbon-cell-based brain.

Will the new entity be capable of spiritual experiences? Because its brain processes are effectively identical, its behavior will be comparable to that of the person it is based on. So it will certainly claim to have the full range of emotional and spiritual experiences that a person claims to have.

No objective test can absolutely determine consciousness. We cannot objectively measure subjective experience (this has to do with the very nature of the concepts "objective" and "subjective"). We can measure only correlates of it, such as behavior. The new entities will appear to be conscious, and whether or not they actually are will not affect their behavior. Just as we debate today the consciousness of nonhuman entities such as animals, we will surely debate the potential consciousness of nonbiological intelligent entities. From a practical perspective, we will accept their claims. They'll get mad if we don't.

Before the next century is over, the Law of Accelerating Returns tells us, Earth's technology-creating species—us—will merge with our own technology. And when that happens, we might ask: What is the difference between a human brain enhanced a millionfold by neural implants and a nonbiological intelligence based on the reverse-engineering of the human brain that is subsequently enhanced and expanded?

The engine of evolution used its innovation from one period (humans) to create the next (intelligent machines). The subsequent milestone will be for the machines to create their own next generation without human intervention.

An evolutionary process accelerates because it builds on its own means for further evolution. Humans have beaten evolution. We are creating intelligent entities in considerably less time than it took the evolutionary process to create us. Human intelligence—a product of evolution—has transcended it. So, too, the intelligence that we are now creating in computers will soon exceed the intelligence of its creators.

RAY KURZWEIL is CEO of Kurzweil Technologies, Inc. He led teams that built a pioneering print-to-speech reading machine, the first omni-font ("any" font) optical-character-recognition system, the first text-to-speech synthesizer, the first music synthesizer capable of re-creating the grand piano and the first commercially marketed large-vocabulary speech-recognition system.
Originally published in *Scientific American Presents*, Vol. 10, No.3, Fall 1999.

18

Controlling Robots with the Mind

People with nerve or limb injuries may one day be able to command
wheelchairs, prosthetics and even paralyzed arms and legs by
"thinking them through" the motions

MIGUEL A. L. NICOLELIS AND JOHN K. CHAPIN

Belle, our tiny owl monkey, was seated in her special chair inside a soundproof
chamber at our Duke University laboratory. Her right hand grasped a joystick as
she watched a horizontal series of lights on a display panel. She knew that if a
light suddenly shone and she moved the joystick left or right to correspond to its
position, a dispenser would send a drop of fruit juice into her mouth. She loved
to play this game. And she was good at it.

Belle wore a cap glued to her head. Under it were four plastic connectors.
The connectors fed arrays of microwires—each wire finer than the finest sew-
ing thread—into different regions of Belle's motor cortex, the brain tissue that
plans movements and sends instructions for enacting the plans to nerve cells in
the spinal cord. Each of the 100 microwires lay beside a single motor neuron.
When a neuron produced an electrical discharge—an "action potential"—the
adjacent microwire would capture the current and send it up through a small
wiring bundle that ran from Belle's cap to a box of electronics on a table next to
the booth. The box, in turn, was linked to two computers, one next door and the
other half a country away.

In a crowded room across the hall, members of our research team were getting anxious. After months of hard work, we were about to test the idea that
we could reliably translate the raw electrical activity in a living being's brain—
Belle's mere thoughts—into signals that could direct the actions of a robot. Unknown to Belle on this spring afternoon in 2000, we had assembled a multijointed robot arm in this room, away from her view, that she would control for

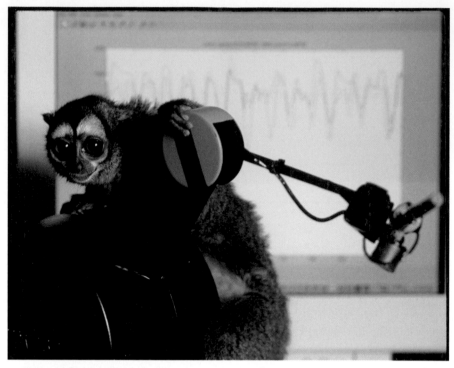

Belle the owl monkey

the first time. As soon as Belle's brain sensed a lit spot on the panel, electronics in the box running two real-time mathematical models would rapidly analyze the tiny action potentials produced by her brain cells. Our lab computer would convert the electrical patterns into instructions that would direct the robot arm. Six hundred miles north, in Cambridge, Mass., a different computer would produce the same actions in another robot arm, built by Mandayam A. Srinivasan, head of the Laboratory for Human and Machine Haptics (the Touch Lab) at the Massachusetts Institute of Technology. At least, that was the plan.

If we had done everything correctly, the two robot arms would behave as Belle's arm did, at exactly the same time. We would have to translate her neuronal activity into robot commands in just 300 milliseconds—the natural delay between the time Belle's motor cortex planned how she should move her limb and the moment it sent the instructions to her muscles. If the brain of a living creature could accurately control two dissimilar robot arms—despite the signal noise and transmission delays inherent in our lab network and the error-prone

Internet—perhaps it could someday control a mechanical device or actual limbs in ways that would be truly helpful to people.

Finally the moment came. We randomly switched on lights in front of Belle, and she immediately moved her joystick back and forth to correspond to them. Our robot arm moved similarly to Belle's real arm. So did Srinivasan's. Belle and the robots moved in synchrony, like dancers choreographed by the electrical impulses sparking in Belle's mind. Amid the loud celebration that erupted in Durham, N.C., and Cambridge, we could not help thinking that this was only the beginning of a promising journey.

In the two years since that day, our labs and several others have advanced neuroscience, computer science, microelectronics and robotics to create ways for rats, monkeys and eventually humans to control mechanical and electronic machines purely by "thinking through," or imagining, the motions. Our immediate goal is to help a person who has been paralyzed by a neurological disorder or spinal cord injury, but whose motor cortex is spared, to operate a wheelchair or a robotic limb. Someday the research could also help such a patient regain control over a natural arm or leg, with the aid of wireless communication between implants in the brain and the limb. And it could lead to devices that restore or augment other motor, sensory or cognitive functions.

The big question is, of course, whether we can make a practical, reliable system. Doctors have no means by which to repair spinal cord breaks or damaged brains. In the distant future, neuroscientists may be able to regenerate injured neurons or program stem cells (those capable of differentiating into various cell types) to take their place. But in the near future, brain-machine interfaces (BMIs), or neuroprostheses, are a more viable option for restoring motor function. Success this summer with macaque monkeys that completed different tasks than those we asked of Belle has gotten us even closer to this goal.

FROM THEORY TO PRACTICE

Recent advances in brain-machine interfaces are grounded in part on discoveries made about 20 years ago. In the early 1980s Apostolos P. Georgopoulos of Johns Hopkins University recorded the electrical activity of single motor-cortex neurons in macaque monkeys. He found that the nerve cells typically reacted most strongly when a monkey moved its arm in a certain direction. Yet when the arm moved at an angle away from a cell's preferred direction, the neuron's activity didn't cease; it diminished in proportion to the cosine of that angle. The

OVERVIEW/BRAIN INTERFACES

- Rats and monkeys whose brains have been wired to a computer have successfully controlled levers and robot arms by imagining their own limb either pressing a bar or manipulating a joystick.
- These feats have been made possible by advances in microwires that can be implanted in the motor cortex and by the development of algorithms that translate the electrical activity of brain neurons into commands able to control mechanical devices.
- Human trials of sophisticated brain-machine interfaces are far off, but the technology could eventually help people who have lost an arm to control a robotic replacement with their mind or help patients with a spinal cord injury regain control of a paralyzed limb.

finding showed that motor neurons were broadly tuned to a range of motion and that the brain most likely relied on the collective activity of dispersed populations of single neurons to generate a motor command.

There were caveats, however. Georgopoulos had recorded the activity of single neurons one at a time and from only one motor area. This approach left unproved the underlying hypothesis that some kind of coding scheme emerges from the simultaneous activity of many neurons distributed across multiple cortical areas. Scientists knew that the frontal and parietal lobes—in the forward and rear parts of the brain, respectively—interacted to plan and generate motor commands. But technological bottlenecks prevented neurophysiologists from making widespread recordings at once. Furthermore, most scientists believed that by cataloguing the properties of neurons one at a time, they could build a comprehensive map of how the brain works—as if charting the properties of individual trees could unveil the ecological structure of an entire forest!

Fortunately, not everyone agreed. When the two of us met 14 years ago at Hahnemann University, we discussed the challenge of simultaneously recording many single neurons. By 1993 technological breakthroughs we had made allowed us to record 48 neurons spread across five structures that form a rat's sensorimotor system—the brain regions that perceive and use sensory information to direct movements.

Crucial to our success back then—and since—were new electrode arrays containing Teflon-coated stainless-steel microwires that could be implanted in an animal's brain. Neurophysiologists had used standard electrodes that resemble rigid needles to record single neurons. These classic electrodes worked well but only for a few hours, because cellular compounds collected around the electrodes' tips and eventually insulated them from the current. Furthermore, as the subject's brain moved slightly during normal activity, the stiff pins damaged neurons. The microwires we devised in our lab (later produced by NBLabs in Denison, Tex.) had blunter tips, about 50 microns in diameter, and were much more flexible. Cellular substances did not seal off the ends, and the flexibility greatly reduced neuron damage. These properties enabled us to produce recordings for months on end, and having tools for reliable recording allowed us to begin developing systems for translating brain signals into commands that could control a mechanical device.

With electrical engineer Harvey Wiggins, now president of Plexon in Dallas, and with Donald J. Woodward and Samuel A. Deadwyler of Wake Forest University School of Medicine, we devised a small "Harvey box" of custom electronics, like the one next to Belle's booth. It was the first hardware that could properly sample, filter and amplify neural signals from many electrodes. Special software allowed us to discriminate electrical activity from up to four single neurons per microwire by identifying unique features of each cell's electrical discharge.

A RAT'S BRAIN CONTROLS A LEVER

In our next experiments at Hahnemann in the mid-1990s, we taught a rat in a cage to control a lever with its mind. First we trained it to press a bar with its forelimb. The bar was electronically connected to a lever outside the cage. When the rat pressed the bar, the outside lever tipped down to a chute and delivered a drop of water it could drink.

We fitted the rat's head with a small version of the brain-machine interface Belle would later use. Every time the rat commanded its forelimb to press the bar, we simultaneously recorded the action potentials produced by 46 neurons. We had programmed resistors in a so-called integrator, which weighted and processed data from the neurons to generate a single analog output that predicted very well the trajectory of the rat's forelimb. We linked this integrator to the robot lever's controller so that it could command the lever.

Once the rat had gotten used to pressing the bar for water, we disconnected the bar from the lever. The rat pressed the bar, but the lever remained still. Frustrated, it began to press the bar repeatedly, to no avail. But one time, the lever tipped and delivered the water. The rat didn't know it, but its 46 neurons had expressed the same firing pattern they had in earlier trials when the bar still worked. That pattern prompted the integrator to put the lever in motion.

After several hours the rat realized it no longer needed to press the bar. If it just looked at the bar and imagined its forelimb pressing it, its neurons could still express the firing pattern that our brain-machine interface would interpret as motor commands to move the lever. Over time, four of six rats succeeded in this task. They learned that they had to "think through" the motion of pressing the bar. This is not as mystical at it might sound; right now you can imagine reaching out to grasp an object near you—without doing so. In similar fashion, a person with an injured or severed limb might learn to control a robot arm joined to a shoulder.

A MONKEY'S BRAIN CONTROLS A ROBOT ARM

We were thrilled with our rats' success. It inspired us to move forward, to try to reproduce in a robotic limb the three-dimensional arm movements made by monkeys—animals with brains far more similar to those of humans. As a first step, we had to devise technology for predicting how the monkeys intended to move their natural arms.

At this time, one of us (Nicolelis) moved to Duke and established a neuro-physiology laboratory there. Together we built an interface to simultaneously monitor close to 100 neurons, distributed across the frontal and parietal lobes. We proceeded to try it with several owl monkeys. We chose owl monkeys because their motor cortical areas are located on the surface of their smooth brain, a configuration that minimizes the surgical difficulty of implanting microwire arrays. The microwire arrays allowed us to record the action potentials in each creature's brain for several months.

In our first experiments, we required owl monkeys, including Belle, to move a joystick left or right after seeing a light appear on the left or right side of a video screen. We later sat them in a chair facing an opaque barrier. When we lifted the barrier they saw a piece of fruit on a tray. The monkeys had to reach out and grab the fruit, bring it to their mouth and place their hand back down. We measured the position of each monkey's wrist by attaching fiber-optic sensors to it, which defined the wrist's trajectory.

Further analysis revealed that a simple linear summation of the electrical activity of cortical motor neurons predicted very well the position of an animal's hand a few hundred milliseconds ahead of time. This discovery was made by Johan Wessberg of Duke, now at the Gothenburg University in Sweden. The main trick was for the computer to continuously combine neuronal activity produced as far back in time as one second to best predict movements in real time.

As our scientific work proceeded, we acquired a more advanced Harvey box from Plexon. Using it and some custom, real-time algorithms, our computer sampled and integrated the action potentials every 50 to 100 milliseconds. Software translated the output into instructions that could direct the actions of a robot arm in three-dimensional space. Only then did we try to use a BMI to control a robotic device. As we watched our multijointed robot arm accurately mimic Belle's arm movements on that inspiring afternoon in 2000, it was difficult not to ponder the implausibility of it all. Only 50 to 100 neurons randomly sampled from tens of millions were doing the needed work.

Later mathematical analyses revealed that the accuracy of the robot movements was roughly proportional to the number of neurons recorded, but this linear relation began to taper off as the number increased. By sampling 100 neurons we could create robot hand trajectories that were about 70 percent similar to those the monkeys produced. Further analysis estimated that to achieve 95 percent accuracy in the prediction of one-dimensional hand movements, as few as 500 to 700 neurons would suffice, depending on which brain regions we sampled. We are now calculating the number of neurons that would be needed for highly accurate three-dimensional movements. We suspect the total will again be in the hundreds, not thousands.

These results suggest that within each cortical area, the "message" defining a given hand movement is widely disseminated. This decentralization is extremely beneficial to the animal: in case of injury, the animal can fall back on a huge reservoir of redundancy. For us researchers, it means that a BMI neuroprosthesis for severely paralyzed patients may require sampling smaller populations of neurons than was once anticipated.

We continued working with Belle and our other monkeys after Belle's successful experiment. We found that as the animals perfected their tasks, the properties of their neurons changed—over several days or even within a daily two-hour recording session. The contribution of individual neurons varied over time. To cope with this "motor learning," we added a simple routine that enabled our model to reassess periodically the contribution of each neuron. Brain cells that ceased to influence the predictions significantly were dropped from the model, and those that became better predictors were added. In essence, we

BELLE'S 600-MILE REACH

On the day Belle first moved a multijointed robot arm with her thoughts, she wore a cap glued to her head. Beneath the cap, each of four plastic connectors fed an array of fine microwires into her cortex (a). As Belle saw lights shine suddenly and decided to move a joystick left or right to correspond to them, the microwires detected electrical signals produced by activated neurons in her cortex and relayed the signals to a "Harvey box" of electronics.

The box collected, filtered and amplified the signals and relayed them to a server computer in a room next door. The signals received by the box can

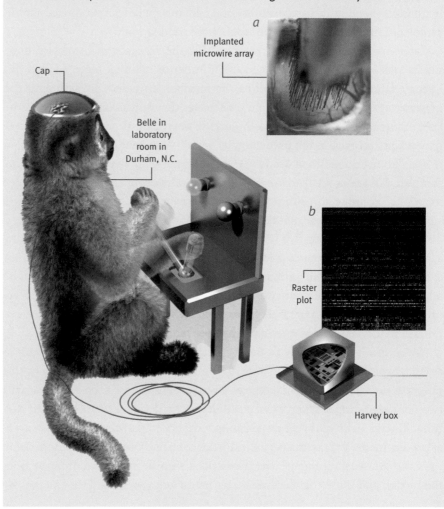

a

Implanted
microwire array

Cap

Belle in
laboratory
room in
Durham, N.C.

b

Raster
plot

Harvey box

be displayed as a raster plot (b); each row represents the activity of a single neuron recorded over time, and each color bar indicates that the neuron was firing at a given moment.

The computer, in turn, predicted the trajectory that Belle's arm would take (c) and converted that information into commands for producing the same motion in a robot arm. Then the computer sent commands to another computer that operated a robot arm in a room across the hall. At the same time, it sent commands from our laboratory in Durham, N.C., to another robot in a laboratory hundreds of miles away. In response, both robot arms moved in synchrony with Belle's own limb. —M.A.L.N. and J.K.C.

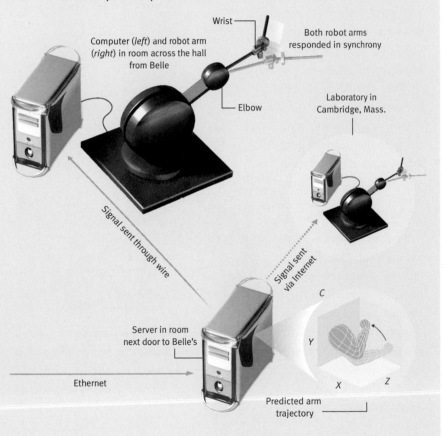

Computer (*left*) and robot arm (*right*) in room across the hall from Belle

Wrist

Both robot arms responded in synchrony

Elbow

Laboratory in Cambridge, Mass.

Signal sent through wire

Signal sent via Internet

Server in room next door to Belle's

Ethernet

C

Y

X

Z

Predicted arm trajectory

A VISION OF THE FUTURE

A brain-machine interface might someday help a patient whose limbs have been paralyzed by a spine injury. Tiny arrays of microwires implanted in multiple motor cortex areas of the brain would be wired to a neurochip in the skull. As the person imagined her paralyzed arm moving in a particular way, such as reaching out for food on a table, the chip would convert the thoughts into a train of radio-frequency signals and send them wirelessly to a small battery-operated "backpack" computer hanging from the chair.

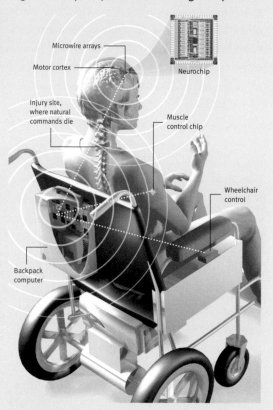

The computer would convert the signals into motor commands and dispatch them, again wirelessly, to a different chip implanted in the person's arm. This second chip would stimulate nerves needed to move the arm muscles in the desired fashion. Alternatively, the backpack computer could control the wheelchair's motor and steering directly, as the person envisioned where she wanted the chair to roll. Or the computer could send signals to a robotic arm if a natural arm were missing or to a robot arm mounted on a chair. Patrick D. Wolf of Duke University has built a prototype neurochip and backpack, as envisioned here.
—M.A.L.N. and J.K.C.

designed a way to extract from the brain a neural output for hand trajectory. This coding, plus our ability to measure neurons reliably over time, allowed our BMI to represent Belle's intended movements accurately for several months. We could have continued, but we had the data we needed.

It is important to note that the gradual changing of neuronal electrical activity helps to give the brain its plasticity. The number of action potentials a neuron generates before a given movement changes as the animal undergoes more experiences. Yet the dynamic revision of neuronal properties does not represent an impediment for practical BMIs. The beauty of a distributed neural output is that it does not rely on a small group of neurons. If a BMI can maintain viable recordings from hundreds to thousands of single neurons for months to years and utilize models that can learn, it can handle evolving neurons, neuronal death and even degradation in electrode-recording capabilities.

EXPLOITING SENSORY FEEDBACK

Belle proved that a BMI can work for a primate brain. But could we adapt the interface to more-complex brains? In May 2001 we began studies with three macaque monkeys at Duke. Their brains contain deep furrows and convolutions that resemble those of the human brain.

We employed the same BMI used for Belle, with one fundamental addition: now the monkeys could exploit visual feedback to judge for themselves how well the BMI could mimic their hand movements. We let the macaques move a joystick in random directions, driving a cursor across a computer screen. Suddenly a round target would appear somewhere on the screen. To receive a sip of fruit juice, the monkey had to position the cursor quickly inside the target—within 0.5 seconds—by rapidly manipulating the joystick.

The first macaque to master this task was Aurora, an elegant female who clearly enjoyed showing off that she could hit the target more than 90 percent of the time. For a year, our postdoctoral fellows Roy Crist and José Carmena recorded the activity of up to 92 neurons in five frontal and parietal areas of Aurora's cortex.

Once Aurora commanded the game, we started playing a trick on her. In about 30 percent of the trials we disabled the connection between the joystick and the cursor. To move the cursor quickly within the target, Aurora had to rely solely on her brain activity, processed by our BMI. After being puzzled, Aurora gradually altered her strategy. Although she continued to make hand movements, after a few days she learned she could control the cursor 100 percent of the time with her brain alone. In a few trials each day during the ensuing weeks Aurora didn't even bother to move her hand; she moved the cursor by just thinking about the trajectory it should take.

That was not all. Because Aurora could see her performance on the screen, the BMI made better and better predictions even though it was recording the same neurons. Although much more analysis is required to understand this result, one explanation is that the visual feedback helped Aurora to maximize the BMI's reaction to both brain and machine learning. If this proves true, visual or other sensory feedback could allow people to improve the performance of their own BMIs.

We observed another encouraging result. At this writing, it has been a year since we implanted the microwires in Aurora's brain, and we continue to record 60 to 70 neurons daily. This extended success indicates that even in a primate with a convoluted brain, our microwire arrays can provide long-term, high-quality, multichannel signals. Although this sample is down from the original 92 neurons, Aurora's performance with the BMI remains at the highest levels she has achieved.

We will make Aurora's tasks more challenging. In May we began modifying the BMI to give her tactile feedback for new experiments that are now beginning. The BMI will control a nearby robot arm fitted with a gripper that simulates a grasping hand. Force sensors will indicate when the gripper encounters an object and how much force is required to hold it. Tactile feedback—is the object heavy or light, slick or sticky?—will be delivered to a patch on Aurora's skin embedded with small vibrators. Variations in the vibration frequencies should help Aurora figure out how much force the robot arm should apply to, say, pick up a piece of fruit, and to hold it as the robot brings it back to her. This experiment might give us the most concrete evidence yet that a person suffering from severe paralysis could regain basic arm movements through an implant in the brain that communicates over wires, or wirelessly, with signal generators embedded in a limb.

If visual and tactile sensations mimic the information that usually flows between Aurora's own arm and brain, long-term interaction with a BMI could possibly stimulate her brain to incorporate the robot into its representations of her body-schema known to exist in most brain regions. In other words, Aurora's brain might represent this artificial device as another part of her body. Neuronal tissue in her brain might even dedicate itself to operating the robot arm and interpreting its feedback.

To test whether this hypothesis has merit, we plan to conduct experiments

STOPPING SEIZURES

Recent experiments suggest that brain-machine interfaces could one day help prevent brain seizures in people who suffer from severe chronic epilepsy, which causes dozens of seizures a day. The condition ruins a patient's quality of life and can lead to permanent brain damage. To make matters worse, patients usually become unresponsive to traditional drug therapy.

A BMI for seizure control would function somewhat like a heart pacemaker. It would continuously monitor the brain's electrical activity for patterns that indicate an imminent attack. If the BMI sensed such a pattern, it would deliver an electrical stimulus to the brain or a peripheral nerve that would quench the rising storm or trigger the release of antiepileptic medication.

At Duke we demonstrated the feasibility of this concept in collaboration with Erika E. Fanselow, now at Brown University, and Ashlan P. Reid, now at the University of Pennsylvania. We implanted a BMI with arrays of microwires in rats given PTZ, a drug that induces repetitive mild epilepsy. When a seizure starts, cortical neurons begin firing together in highly synchronized bursts. When the "brain pacemaker" detected this pattern, it triggered the electrical stimulation of the large trigeminal cranial nerve. The brief stimulus disrupted the epileptic activity quickly and efficiently, without damaging the nerve, and reduced the occurrence and duration of seizures.

—M.A.L.N. and J.K.C.

like those done with Aurora, except that an animal's arm will be temporarily anesthetized, thereby removing any natural feedback information. We predict that after a transition period, the primate will be able to interact with the BMI just fine. If the animal's brain does meld the robot arm into its body representations, it is reasonable to expect that a paraplegic's brain would do the same, rededicating neurons that once served a natural limb to the operation of an artificial one.

Each advance shows how plastic the brain is. Yet there will always be limits. It is unlikely, for example, that a stroke victim could gain full control over a robot limb. Stroke damage is usually widespread and involves so much of the brain's white matter—the fibers that allow brain regions to communicate—that the destruction overwhelms the brain's plastic capabilities. This is why stroke victims who lose control of uninjured limbs rarely regain it.

REALITY CHECK

Good news notwithstanding, we researchers must be very cautious about offering false hope to people with serious disabilities. We must still overcome many hurdles before BMIs can be considered safe, reliable and efficient therapeutic options. We have to demonstrate in clinical trials that a proposed BMI will offer much greater well-being while posing no risk of added neurological damage.

Surgical implantation of electrode arrays will always be of medical concern, for instance. Investigators need to evaluate whether highly dense microwire arrays can provide viable recordings without causing tissue damage or infection in humans. Progress toward dense arrays is already under way. Duke electronics technician Gary Lehew has designed ways to increase significantly the number of microwires mounted in an array that is light and easy to implant. We can now implant multiple arrays, each of which has up to 160 microwires and measures five by eight millimeters, smaller than a pinky fingernail. We recently implanted 704 microwires across eight cortical areas in a macaque and recorded 318 neurons simultaneously.

In addition, considerable miniaturization of electronics and batteries must occur. We have begun collaborating with José Carlos Príncipe of the University of Florida to craft implantable microelectronics that will embed in hardware the neuronal pattern recognition we now do with software, thereby eventually freeing the BMI from a computer. These microchips will thus have to send wireless control data to robotic actuators. Working with Patrick D. Wolf's lab at Duke, we have built the first wireless "neurochip" and beta-tested it with Aurora. Seeing streams of neural activity flash on a laptop many meters away from Aurora—broadcast via the first wireless connection between a primate's brain and a computer—was a delight.

More and more scientists are embracing the vision that BMIs can help people in need. In the past year, several traditional neurological laboratories have begun to pursue neuroprosthetic devices. Preliminary results from Arizona State University, Brown University and the California Institute of Technology have recently appeared. Some of the studies provide independent confirmation of the rat and monkey studies we have done. Researchers at Arizona State basically reproduced our 3-D approach in owl monkeys and showed that it can work in rhesus monkeys too. Scientists at Brown enabled a rhesus macaque monkey to move a cursor around a computer screen. Both groups recorded 10 to 20 neurons or so per animal. Their success further demonstrates that this new field is progressing nicely.

The most useful BMIs will exploit hundreds to a few thousand single neurons distributed over multiple motor regions in the frontal and parietal lobes. Those that record only a small number of neurons (say, 30 or fewer) from a single cortical area would never provide clinical help, because they would lack the excess capacity required to adapt to neuronal loss or changes in neuronal responsiveness. The other extreme—recording millions of neurons using large electrodes—would most likely not work either, because it might be too invasive.

Noninvasive methods, though promising for some therapies, will probably be of limited use for controlling prostheses with thoughts. Scalp recording, called electroencephalography (EEG), is a noninvasive technique that can drive a different kind of brain-machine interface, however. Niels Birbaumer of the University of Tübingen in Germany has successfully used EEG recordings and a computer interface to help patients paralyzed by severe neurological disorders learn how to modulate their EEG activity to select letters on a computer screen, so they can write messages. The process is time-consuming but offers the only way for these people to communicate with the world. Yet EEG signals cannot be used directly for limb prostheses, because they depict the average electrical activity of broad populations of neurons; it is difficult to extract from them the fine variations needed to encode precise arm and hand movements.

Despite the remaining hurdles, we have plenty of reasons to be optimistic. Although it may be a decade before we witness the operation of the first human neuroprosthesis, all the amazing possibilities crossed our minds that afternoon in Durham as we watched the activity of Belle's neurons flashing on a computer monitor. We will always remember our sense of awe as we eavesdropped on the processes by which the primate brain generates a thought. Belle's thought to receive her juice was a simple one, but a thought it was, and it commanded the outside world to achieve her very real goal.

MORE TO EXPLORE

Chapin, J.K., and K.A. Moxon, eds. 2001. *Neural prostheses for restoration of sensory and motor function*. CRC Press.

Chapin, J.K., K.A. Moxon, R.S. Markowitz, and M.A.L. Nicolelis. 1999. Real-time control of a robot arm using simultaneously recorded neurons in the motor cortex. *Nature Neuroscience* [JM3]2:664–670.

Nicolelis, M.A.L. 2001. Action from thoughts. *Nature* 409 (January 18): 403–407.

———, ed. 2001. Advancesin neural population coding. *Progress in Brain Research* 130. Elsevier.

Wessberg, J.,C.R. Stambaugh, J.D. Kralik, P.D. Beck, J.K. Chapin, J. Kim, S.J. Biggs, M.A. Srinivasan, and M.A.L. Nicolelis. 2000. Real-time prediction of hand trajectory by ensembles of cortical neurons in primates. *Nature* 408 (November 16): 361–365.

MIGUEL A. L. NICOLELIS and JOHN K. CHAPIN have collaborated for more than a decade. Nicolelis, a native of Brazil, received his M.D. and Ph.D. in neurophysiology from the University of São Paulo. After postdoctoral work at Hahnemann University, he joined Duke University, where he now co-directs the Center for Neuroengineering and is professor of neurobiology, biomedical engineering, and psychological and brain sciences. Chapin received his Ph.D. in neurophysiology from the University of Rochester and has held faculty positions at the University of Texas and the MCP Hahnemann University School of Medicine (now Drexel University College of Medicine). He is currently professor of physiology and pharmacology at the State University of New York Downstate Medical Center.

Originally published in *Scientific American*, Vol. 287, No.4, October 2002.

19

Thinking Out Loud

Thought-deciphering systems are enabling paralyzed people to communicate—
and someday may let them control wheelchairs, prosthetics and even their own
muscles

NICOLA NEUMANN AND NIELS BIRBAUMER

Consider the plight of Hans-Peter Salzmann. The 49-year-old former lawyer is confined to a wheelchair and cannot eat or breathe on his own. For the past 15 years he has been suffering from amyotrophic lateral sclerosis, also known as ALS or Lou Gehrig's disease, an incurable degenerative disease of the nerve cells that breaks down the entire voluntary motor system. To spell out words, Salzmann blinks his left eye to choose letters from a printed list on a board, a tedious process that requires an experienced interpreter. Sometimes his eyelid is too weak to make the selections. Ultimatcly, people with ALS, brain stem stroke or other illnesses may lose all ability to move—becoming a functioning mind "locked in" an immobile body.

Now technologies called brain-computer interfaces, which read aspects of brain activity and react to them, are offering patients such as Salzmann ways to continue to express themselves despite their disabilities. The systems enable a person to use his mind to guide cursors on a screen for communication. Someday they may lead to mental command of environmental-control devices in a home, "smart" wheelchairs and prosthetics.

READING SIGNALS

Since 1929, when Hans Berger first described the electroencephalogram (EEG), an instrument that could read electrical impulses produced by nerve cells, people have speculated that it might be used for communication and control. Electrodes attached to the scalp measure the voltage difference, or potential, between two points in the brain. A few laboratories created prototype brain-computer interfaces in the 1980s and have been refining them since then.

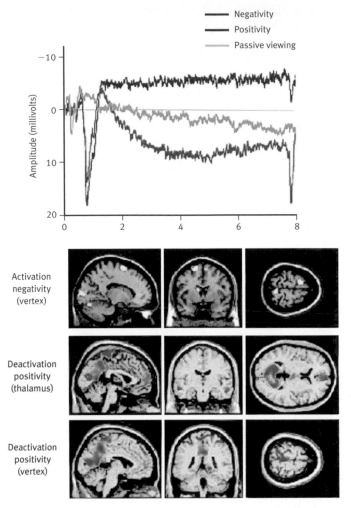

Brain waves displayed on a monitor during training give patients feedback. Imaging shows corresponding areas of activity.

One such unit, developed by one of us (Birbaumer) and his colleagues at the Tübingen Institute of Medical Psychology and Behavioral Neurobiology in Germany, is the Thought Translation Device, which works on the principle of biofeedback. Since 1996 Salzmann and 10 other paralyzed patients around the world have been testing the translator. Using the device to report the status of brain waves called the slow cortical potential (SCP), the patient learns to manage a normally imperceptible physiological occurrence. Unlike the millisecond pulses typically measured by an EEG, SCPs build up over several seconds. This relatively slow speed makes SCPs the easiest brain waves to detect by outside means and to be influenced by the patient himself. These brain waves are not necessarily connected to concrete actions or feelings; rather they correspond to the general state of activity in the brain.

Electrodes attached to the top of the head record the brain waves, which are amplified, transmitted to a control processor with an analog-digital transformer card and then sent to a notebook computer. On the monitor, the patient can observe the progression of his SCP as a moving cursor. When the machine reads an electrically negative potential, the cursor rises in response; a positive potential drops it downward. The challenge is to learn how to deliberately move the cursor into either of two goals, at the top or bottom of the computer screen. When the patient scores a goal, he sees one point added to his total and a smiley face appears—this simple reward has been proved to increase the success rate. In each session, the person repeats the activity several hundred times. After a few weeks, many subjects are able to steer the cursor correctly some 70 to 80 percent of the time.

When asked how they control their SCP, patients offer various answers. Salzmann reports that when he wants to push the cursor upward, he attempts to think of nothing at all. To force the cursor downward, he imagines a situation that involves anticipation and release—such as a traffic light turning from red to green or a sprinter starting a race. Others may think of specific words or previous tasks that required concentration. Some do not even think about anything in particular, but rather they move the cursor much as they would move their own limbs, without making any specific associations.

Once a patient masters the cursor, he can use this skill to select letters from the lower part of the screen. If the one he wants is not there, the patient signifies this lack by guiding the cursor away from the field of letters. Every time he refuses a set, a new one appears, which he then either accepts or rejects. Each selected group is cut in half until only the desired letter remains. The system also contains a list of common words from which to choose.

Even with the aid of this tool, it may take most of an hour to pick 100 characters or several days to compose a full letter. Still, the ability to correspond without the need for an interpreter has enabled patients to regain a very important part of their private lives.

LIVE CHAT

Enabling spontaneous conversation would be preferable, but that would require extensive improvements in the technology. Toward this end, our research group at Tübingen began work in early 2000 with Jonathan Wolpaw and his colleagues at the Wadsworth Center of the New York State Department of Health and others. Together we created the BCI2000, a flexible and universal platform on which new brain-wave technologies could be tested.

Scientists on Wolpaw's team work not with SCP but with mu waves, which have frequencies between 8 and 12 hertz, and beta rhythms, which have about double that frequency range. The system enables a person to move the cursor up or down by raising or lowering the amplitude of the mu or beta rhythms. These oscillations occur when a subject uses motor skills for movement or when he simply imagines such movement; a typical strategy is to imagine lifting or lowering a hand or other body part.

With the combined brain-computer interface, paralyzed patients can select a signal that they can operate with the most accuracy. We also have new interpretational programs that allow for differentiation between more than two cursor states—for example, the cursor could also be moved to the left and right as well as up and down.

Another type of brain-wave sensor integrated into the BCI2000 device is for detection of the P300 potential, a brief voltage increase that peaks about 300 milliseconds after the brain registers the onset of a surprising event. Emanuel

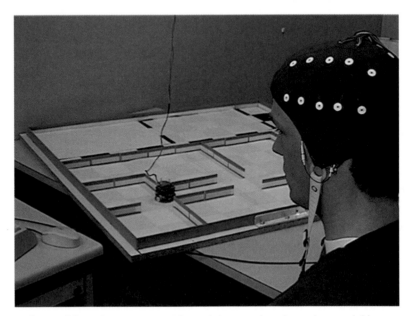

EEG readings of thought patterns guide a miniature robot through a model house.

Donchin, a psychologist emeritus at the University of Illinois and now at the University of South Florida, has focused on "event-related potentials," particularly on P300.

Donchin's system exploits the fact that the human brain reacts very differently to a novel stimulus among standard ones—in this case, the target letter rather than any of the others. Subjects focus attention on a specific character within a letter matrix. While the separate rows and columns within the matrix light up, one after another, the person must count how many times the desired letter appears. When the brain realizes "There it is!" it generates a P300 potential. A computer program then compares which rows and which columns elicited P300 waves and thereby identifies the desired letter. Students without neurological impairments who participated in tests were able to select up to eight characters per minute with a high level of accuracy. Researchers are currently testing the performance of Donchin's brain-computer interface in paralyzed individuals.

An advantage of this method is that it can recognize thoughts within a predetermined category—in this case, letters of the alphabet—without the chore of learning to regulate brain activity, as SCP detectors require. On the downside, the device cannot make sense of anything going on in the brain outside this predetermined category.

Taking the idea of mental control a step beyond the cursor, José del R. Millán and his colleagues at the Dalle Molle Institute for Perceptual Artificial Intelligence in Martigny, Switzerland, have developed an interface that analyzes overall EEG signals at eight locations on the skull. It captures the different types of patterns produced by thinking about very different things. Using a neural network algorithm, a computer learns to distinguish among three types of such thoughts. It is then able to perform a programmed command based on the mental pattern that it detects. In experiments, healthy individuals learn to direct a small wheeled robot (a stand-in for a smart wheelchair).

LISTENING IN

The systems described here all rely on EEG measurements of the activities of millions of nerve cells—thereby making these approaches rather imprecise. The process could be compared to trying to hear the conversation between two individuals sitting in a packed sports stadium with the use of a directional microphone located in the parking lot. Wouldn't it be much more practical to listen in to conversations among nerve cells from closer range?

Miguel A. L. Nicolelis and his colleagues at Duke University are attempting to create exactly this kind of situation. In 2001 they implanted so-called multi-microelectrode arrays in various regions of the motor cortex in monkeys. The monkeys used a joystick to guide a cursor on a computer screen into a goal, while Nicolelis measured the signals related to this motion in up to 92 motor neurons using the implanted electrode arrays. The monkeys had to carry out the hand motion many times so that Nicolelis's team could calculate a mathematical algorithm that would properly assess the activity of the individual nerve cells. Then the joystick was disabled, and control of the cursor was left to the nerve cell activity alone. The monkeys steered the cursor purely by thought—probably by visualizing its path of movement.

This was a considerable advance, in that the scientists succeeded for the first time ever in translating a neural signal based entirely on a mental visualization into a real, two-dimensional movement. With a setup like this, thought reading could become reality. If a brain activity is measured at the exact moment that a person is having a thought, that same thought could be recognized later, by comparing it with a reading of the identical activity.

But how many of the countless ideas that go through our heads every day could be linked with the corresponding activation patterns? Basically a thought in the brain is not imaged by the firing of one single nerve cell but rather by means of the activity in entire cell structures. Such a neural network combines the individual aspects of a piece of information into a complete impression. For example, a woman walks into a café and immediately has a hankering for a delicious, steaming cup of cappuccino, just like the one sitting on the counter. This desire is represented by synchronous activities of various nerve cells: some neurons react to the smell of the coffee, others to the color and form of the cup; still others guide the customer's memory to her last mug of cappuccino.

To "measure" these thoughts, it is not enough to simply record which nerve cells are firing together and which electrochemical processes accompany this event. One also has to know what these cells represent for that individual—say, whether the neural impulses in the hippocampus, our brain's memory storage center, stand for a pleasant or an unpleasant past experience with cappuccino. This memory-signal recognition requires registering the activities of millions of individual nerve cells, and such imaging is not yet possible even with the most advanced visual technologies and invasive means of measuring brain activity.

Human tests of such mind readers are far off. Nevertheless, the work of Nicolelis and others points to the promise of the brain-computer interface to eventually enable paralyzed people to control their surroundings, perhaps even their own bodies. Building on that concept, Patrick D. Wolf, also at Duke, built a prototype neurochip and computer "backpack" that might allow a person to move limbs that have been stilled by spinal injury. Tiny arrays in the brain wired to a chip in the skull would convert electrical activity to radio signals, which would be sent wirelessly to the backpack. The processor would forward the signals to chips in limbs to stimulate nerves directly, thereby moving muscles.

Although we have far to go in achieving such empowerment, brain-computer interfaces offer hope for a better life to those who suffer from serious disabilities.

MORE TO EXPLORE

Birbaumer, Niels, et al. 1999. A spelling device for the paralysed. *Nature* 398 (March 25): 297–298.

Neumann, Nicola, et al. 2003. Conscious perception of brain states: Mental strategies for brain-computer communication. *Neuropsychologia* 41 (8): 1028–1036.

Nicolelis, Miguel A. L., and John K. Chapin. 2003. Controlling robots with the mind. *Scientific American* 287 (4): 46–53.

Wickelgren, Ingrid. 2003. Tapping the mind. *Science* 299 (January 24): 496–499.

NICOLA NEUMANN and NIELS BIRBAUMER collaborate on brain-computer interfaces. Neumann is an assistant professor at the Institute of Medical Psychology and Behavioral Neurobiology at the University of Tübingen in Germany. Birbaumer is director of the institute as well as a professor at the Center for Cognitive Neurosciences at the University of Trento in Italy. For his work in helping epileptics stave off impending seizures by controlling their own slow cortical potential (rather than with drugs), Birbaumer won the Leibniz Prize in medicine in 1995. He used the $1.5 million award to research ways to help locked-in patients communicate.
Originally published in *Scientific American Mind*, Vol. 14, No. 5, December 2004.

20

Neuromorphic Microchips

Compact, efficient electronics based on the brain's neural system could yield
implantable silicon retinas to restore vision, as well as robotic eyes and other
smart sensors

KWABENA BOAHEN

When IBM's Deep Blue supercomputer edged out world chess champion Garry
Kasparov during their celebrated match in 1997, it did so by means of sheer
brute force. The machine evaluated some 200 million potential board moves
a second, whereas its flesh-and-blood opponent considered only three each second, at most. But despite Deep Blue's victory, computers are no real competition for the human brain in areas such as vision, hearing, pattern recognition,
and learning. Computers, for instance, cannot match our ability to recognize a
friend from a distance merely by the way he walks. And when it comes to operational efficiency, there is no contest at all. A typical room-size supercomputer
weighs roughly 1,000 times more, occupies 10,000 times more space and consumes a millionfold more power than does the cantaloupe-size lump of neural
tissue that makes up the brain.

How does the brain—which transmits chemical signals between neurons in
a relatively sluggish thousandth of a second—end up performing some tasks
faster and more efficiently than the most powerful digital processors? The
secret appears to reside in how the brain organizes its slow-acting electrical
components.

The brain does not execute coded instructions; instead it activates links, or
synapses, between neurons. Each such activation is equivalent to executing a
digital instruction, so one can compare how many connections a brain activates
every second with the number of instructions a computer executes during the
same time. Synaptic activity is staggering: 10 quadrillion (10^{16}) neural connections a second. It would take one million Intel Pentium-powered computers to
match that rate—plus a few hundred megawatts to juice them up.

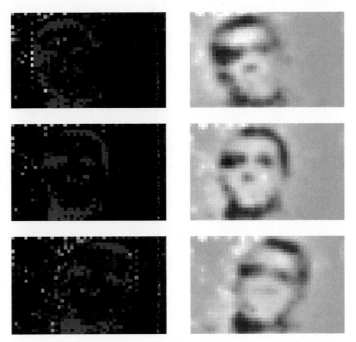

A silicon retina senses the side-to-side head movements of University of Pennsylvania researcher Kareem Zaghloul. The four types of silicon ganglion cells on his Vision chip emulate real retinal cells' ability to preprocess visual information without huge amounts of computation. One class of cells responds to dark areas (red), whereas another reacts to light regions (green). A different set of cells tracks leading edges of objects (yellow) and trailing edges (blue). The gray-scale images, generated by decoding these messages, show what a blind person would see with neuromorphic retinal implants.

Now a small but innovative community of engineers is making significant progress in copying neuronal organization and function. Researchers speak of having "morphed" the structure of neural connections into silicon circuits, creating neuromorphic microchips. If successful, this work could lead to implantable silicon retinas for the blind and sound processors for the deaf that last for 30 years on a single nine-volt battery or to low-cost, highly effective visual, audio or olfactory recognition chips for robots and other smart machines.

Our team at the University of Pennsylvania initially focused on morphing the retina—the half-millimeter-thick sheet of tissue that lines the back of the eye. Comprising five specialized layers of neural cells, the retina "preprocesses" incoming visual images to extract useful information without the need for the brain to expend a great deal of effort. We chose the retina because that sensory

OVERVIEW/INSPIRED BY NATURE
- Today's computers can perform billions of operations per second, but they are still no match for even a young child when it comes to skills such as pattern recognition or visual processing. The human brain is also millions of times more energy-efficient and far more compact than a typical personal computer.
- Neuromorphic microchips, which take cues from neural structure, have already demonstrated impressive power reductions. Their efficiency may make it possible to develop fully implantable artificial retinas for people afflicted by certain types of blindness, as well as better electronic sensors.
- Someday neuromorphic chips could even replicate the self-growing connections the brain uses to achieve its amazing functional capabilities.

system has been well documented by anatomists. We then progressed to morphing the developmental machinery that builds these biological circuits—a process we call metamorphing.

NEUROMORPHING THE RETINA

The nearly one million ganglion cells in the retina compare visual signals received from groups of half a dozen to several hundred photoreceptors, with each group interpreting what is happening in a small portion of the visual field. As features such as light intensity change in a given sector, each ganglion cell transmits pulses of electricity (known as spikes) along the optic nerve to the brain. Each cell fires in proportion to the relative change in light intensity over time or space—not to the absolute input level. So the nerve's sensitivity wanes with growing overall light intensity to accommodate, for example, the five-decade rise in the sky's light levels observed from predawn to high noon.

Misha A. Mahowald, soon after earning her undergraduate biology degree, and Carver Mead, the renowned microelectronics technologist, pioneered efforts to reproduce the retina in silicon at the California Institute of Technology. In their groundbreaking work, Mahowald and Mead reproduced the first three of the retina's five layers electronically ("The Silicon Retina," by Misha A.

NEUROMORPHIC ELECTRONICS RESEARCH GROUPS
Researchers seek to close the efficiency gap between electronic sensors and
the body's neural networks with microchips that emulate the brain. This
work focuses on small sensor systems that can be implanted in the body or
installed in robots.

ORGANIZATION	INVESTIGATORS	PRINCIPAL OBJECTIVES
Johns Hopkins University	Andreas Andreou, Gert Cauwenberghs, Ralph Etienne-Cummings	Battery-powered speech recognizer, rhythm generator for locomotion and camera that extracts object features
ETH Zurich (University of Zurich)	Tobi Delbruck, Shi-Chii Liu, Giacomo Indiveri	Silicon retina and attention chip that automatically select salient regions in a scene
University of Edinburgh	Alan Murray, Alister Hamilton	Artificial noses and automatic odor recognition based on timing of signaling spikes
Georgia Institute of Technology	Steve DeWeerth, Paul Hasler	Coupled rhythm generators that coordinate a multisegmented robot
HKUST, Hong Kong	Bertram Shi	Binocular processor for depth perception and visual tracking
Massachusetts Institute of Technology	Rahul Sarpeshkar	Cochlea-based sound processor for implants for deaf patients
University of Maryland	Timothy Horiuchi	Sonar chip modeled on bat echolocation
University of Arizona	Charles Higgins	Motion-sensing chip based on fly vision

Mahowald and Carver Mead, *Scientific American*, May 1991). Other research-ers, several of whom passed through Mead's Caltech laboratory (the author in-cluded), have morphed succeeding stages of the visual system as well as the auditory system. Kareem Zaghloul morphed all five layers of the retina in 2001 when he was a doctoral student in my lab, making it possible to emulate the visual messages that the ganglion cells, the retina's output neurons, send to the brain. His silicon retina chip, Viso1, replicates responses of the retina's four major types of ganglion cells, which feed into and together make up 90 percent of the optic nerve.

Zaghloul represented the electrical activity of each neuron in the eye's cir-cuitry by an individual voltage output. The voltage controls the current that is conveyed by transistors connected between a given location in the circuit and other points, mimicking how the body modulates the responses of neural syn-apses. Light detected by electronic photosensors affects the voltage in that part of the circuit in a way that is analogous to how it affects a corresponding cell in the retina. And by tiling copies of this basic circuit on his chip, Zaghloul repli-cated the activity in the retina's five cell layers.

The chip emulates the manner in which voltage-activated ion channels cause ganglion cells (and neurons in the rest of the brain) to discharge spikes. To ac-complish this, Zaghloul installed transistors that send current back into the same location in the circuit. When this feedback current arrives, it increases the voltage further, which in turn recruits more feedback current and causes additional amplification. Once a certain initial level is reached, this regenerative effect accelerates, taking the voltage all the way to the highest level, resulting in a spike.

At 60 milliwatts, Zaghloul's neuromorphic chip uses 1,000 times less elec-tricity than a PC. With its low power needs, this silicon retina could pave the way for a total intraocular prosthesis—with camera, processor and stimulator all im-planted inside the eye of a blind person who has retinitis pigmentosa or macular degeneration, diseases that damage photoreceptors but spare the ganglion cells. Retinal prostheses currently being developed, for example, at the University of Southern California, provide what is called phosphene vision—recipients per-ceive the world as a grid of light spots, evoked by stimulating the ganglion cells with microelectrodes implanted inside the eye—and require a wearable com-puter to process images captured by a video camera attached to the patient's glasses. Because the microelectrode array is so small (fewer than 10 pixels by 10 pixels), the patient experiences tunnel vision—head movements are needed to scan scenes.

Alternatively, using the eye itself as the camera would solve the rubbernecking problem, and our chip's 3,600 ganglion-cell outputs should provide near-normal vision. Biocompatible encapsulation materials and stimulation interfaces, however, need further refinement before a high-fidelity prosthesis becomes a reality, maybe by 2010. Better understanding of how various retinal cell types respond to stimulation and how they contribute to perception is also required. In the interim, such neuromorphic chips could find use as sensors in automotive or security applications or in robotic or factory automation systems.

METAMORPHING NEURAL CONNECTIONS

The power savings we attained by morphing the retina were encouraging, a result that started me thinking about how the brain actually achieves high efficiency. Mead was prescient when he recognized two decades ago that even if computing managed to continue along the path of Moore's Law (which states that the number of transistors per square inch on integrated circuits doubles every 24 months), computers as we know them could not reach brainlike efficiency. But how could this be accomplished otherwise? The solution dawned on me eight years ago.

Efficient operation, I realized, comes from the degree to which the hardware is customized for the task at hand. Conventional computers do not allow such adjustments; the software is tailored instead. Today's computers use a few general-purpose tools for every job; software merely changes the order in which the tools are used. In contrast, customizing the hardware is something the brain and neuromorphic chips have in common—they are both programmed at the level of individual connections. They adapt the tool to the specific job. But how does the brain customize itself? If we could translate that mechanism into silicon—metamorphing—we could have our neuromorphic chips modify themselves in the same fashion. Thus, we would not need to painstakingly reverse-engineer the brain's circuits. I started investigating neural development, hoping to learn more about how the body produces exactly the tools it needs.

Building the brain's neural network—a trillion (10^{12}) neurons connected by 10 quadrillion (10^{16}) synapses—is a daunting task. Although human DNA contains the equivalent of a billion bits of information, that amount is not sufficient to specify where all those neurons should go and how they should connect. After employing its genetic information during early development, the brain customizes itself further through internal interactions among neurons and through external interactions with the world outside the body. In other words, sensory neurons wire themselves in response to sensory inputs. The overall rule that

RETINAL NEURONS AND NEUROMORPHIC VISION CHIPS

Biological sensory systems provide compact, energy-efficient models for neuromorphic electronic sensors. Engineers attempting to duplicate the retina in silicon face a tough challenge: the retina is only a half-millimeter thick, weighs half a gram and consumes the equivalent of just one tenth of a watt of power. Recent work at the University of Pennsylvania has yielded a rudimentary silicon retina.

BIOLOGICAL RETINA

The cells in the retina, which are interconnected, extract information from the visual field by engaging in a complex web of excitatory (one-way arrows), inhibitory (circles on a stick), and conductive or bidirectional (two-way arrows) signaling. This circuitry generates the selective responses of the four types of ganglion cells (at bottom) that make up 90 percent of the optic nerve's fibers, which convey visual information to the brain. On (green)

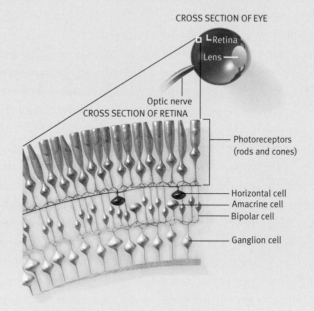

CROSS SECTION OF EYE

Retina
Lens

Optic nerve

CROSS SECTION OF RETINA

Photoreceptors (rods and cones)

Horizontal cell
Amacrine cell
Bipolar cell

Ganglion cell

and Off (red) ganglion cells elevate their firing (spike) rates when the local light intensity is brighter or darker than the surrounding region. Inc (blue) and Dec (yellow) ganglion cells spike when the intensity is increasing or decreasing, respectively.

SILICON RETINA

Neuromorphic circuits emulate the complex interactions that occur among the various retinal cell types by replacing each cell's axons and dendrites (signal pathways) with metal wires and each synapse with a transistor. Permutations of this arrangement produce excitatory and inhibitory interactions that mimic similar communications among neurons. The transistors and the wires that connect them are laid out on silicon chips. Various regions of the chip surface perform the functions of the different cell layers. The large green squares are phototransistors, which transduce light into electricity.

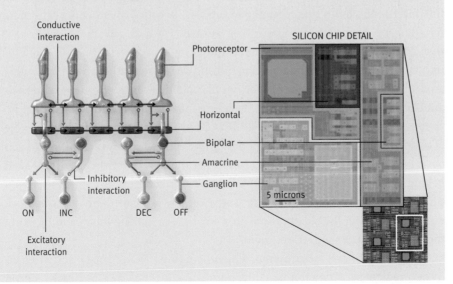

Conductive interaction

Photoreceptor

SILICON CHIP DETAIL

Horizontal

Bipolar

Amacrine

Inhibitory interaction

Ganglion

5 microns

ON   INC   DEC   OFF

Excitatory interaction

regulates this process is deceptively simple: neurons that fire together wire together. That is, out of all the signals that a neuron receives, it accepts those from neurons that are consistently active when it is active, and it ignores the rest.

To learn how one layer of neurons becomes wired to another, neuroscientists have studied the frog's retinotectal projection, which connects its retina to its tectum (the part of the midbrain that processes inputs from sensory organs). They have found that wiring one layer of neurons to another occurs in two stages. A newborn neuron extends projections ("arms") in a multilimbed arbor. The longest arm becomes the axon, the cell's output wire; the rest serve as dendrites, its input wires. The axon then continues to grow, towed by an amoeboid structure at its tip. This growth cone, as scientists call it, senses chemical gradients laid down by trailblazing precursors of neural communication signals, thus guiding the axon to the right street in the tectum's city of cells but not, so to speak, to the right house.

Narrowing the target down to the right house in the tectum requires a second step, but scientists do not understand this process in detail. It is well known, though, that neighboring retinal ganglion cells tend to fire together. This fact led me to speculate that an axon could find its retinal cell neighbors in the tectum by homing in on chemical scents released by active tectal neurons, because its neighbors were most likely at the source of this trail. Once the axon makes contact with the tectal neuron's dendritic arbor, a synapse forms between them and, voilà, the two neurons that fire together are wired together.

In 2001 Brian Taba, a doctoral student in my lab, built a chip modeled on this facet of the brain's developmental process. Because metal wires cannot be rerouted, he decided to reroute spikes instead. He took advantage of the fact that Zaghloul's Vision chip outputs a unique 13-bit address every time one of its 3,600 ganglion cells spikes. Transmitting addresses rather than spikes gets around the limited number of input/output pins that chips have. The addresses are decoded by the receiving chip, which re-creates the spike at the correct location in its silicon neuron mosaic. This technique produces a virtual bundle of axons running between corresponding locations in the two chips—a silicon optic nerve. If we substitute one address with another, we reroute a virtual axon belonging to one neuron (the original address) to another location (the substituted address). We can route these "softwires," as we call them, anywhere we want to by storing the substitutions in a database (a look-up table) and by using the original address to retrieve them.

In Taba's artificial tectum chip, which he named Neurotrope1, softwires activate gradient-sensing circuits (silicon growth cones) as well as nearby silicon neurons, which are situated in the cells of a honeycomb lattice. When active, these silicon neurons release electrical charge into the lattice, which Taba designed to conduct charge like a transistor. Charge diffuses through the lattice much like the chemicals released by tectal cells do through neural tissue. The silicon growth cones sense this simulated diffusing "chemical" and drag their softwires up the gradient—toward the charge's silicon neuron source—by updating the look-up table. Because the charge must be released by the silicon neuron and sensed by the silicon growth cone simultaneously, the softwires end up connecting neurons that are active at the same time. Thus, Neurotrope1 wires together neurons that fire together, as would occur in a real growing axon.

Starting with scrambled wiring between the Visio1 chip and the Neurotrope1 chip, Taba successfully emulated the tendency of neighboring retinal ganglion cells to fire together by activating patches of silicon ganglion cells at random. After stimulating several thousand patches, he observed a dramatic change in the softwiring between the chips. Neighboring artificial ganglion cells now connected to neurons in the silicon tectum that were twice as close as the initial connections. Because of noise and variability, however, the wiring was not perfect: terminals of neighboring cells in the silicon retina did not end up next to one another in the silicon tectum. We wondered how the elaborate wiring patterns thought to underlie biological cortical function arise—and whether we could get further tips from nature to refine our systems.

CORTICAL MAPS

To find out, we had to take a closer look at what neuroscience has learned about connections in the cortex, the brain region responsible for cognition. With an area 16 inches in diameter, the cortex folds like origami paper to fit inside the skull. On this amazing canvas, "maps" of the world outside are drawn during infancy. The best-studied example is what scientists call area V1 (the primary visual cortex), where visual messages from the optic nerve first enter the cortex. Not only are the length and width dimensions of an image mapped onto V1 but also the orientation of the edges of objects therein. As a result, neurons in V1 respond best to edges oriented at a particular angle—vertical lines, horizontal lines, and so forth. The same orientation preferences repeat every millimeter or so, thereby allowing the orientations of edges in different sectors of the visual scene to be detected.

Neurobiologists David H. Hubel and Torsten N. Wiesel, who shared a Nobel Prize in medicine for discovering the V1 map in the 1960s, proposed a wiring diagram for building a visual cortex—one that we found intimidating. According to their model, each cortical cell wires up to two groups of thalamic cells, which act as relays for retinal signals bound for the cortex. One group of thalamic cells should respond to the sensing of dark areas (which we emulate with Visio1's Off cells), whereas the other should react to the sensing of light (like our Visio1's On cells). To make a cortical cell prefer vertical edges, for instance, both groups of cells should be set to lie along a vertical line but should be displaced slightly so the Off cells lie just to the left of the On cells. In that way, a vertical edge of an object in the visual field will activate all the Off cells and all the On cells when it is in the correct position. A horizontal edge, on the other hand, will activate only half the cells in each group. Thus, the cortical cell will receive twice as much input when a vertical edge is present and respond more vigorously.

At first we were daunted by the detail these wiring patterns required. We had to connect each cell according to its orientation preference and then modify these wiring patterns systematically so that orientation preferences changed smoothly, with neighboring cells having similar preferences. As in the cortex, the same orientations would have to be repeated every millimeter, with those silicon cells wired to neighboring locations in the retina. Taba's growth cones certainly could not cope with this complexity. In late 2002 we searched for a way to escape this nightmare altogether. Finally, we found an answer in a five-decade-old experiment.

In the 1950s famed computer scientist Alan M. Turing showed how ordered patterns such as a leopard's spots or a cow's dapples could arise spontaneously from random noise. We hoped we could use a similar technique to create neighboring regions with similar orientation patterns for our chip. Turing's idea, which he tested by running simulations on one of the first electronic computers at the University of Manchester, was that modeled skin cells would secrete "black dye" or "bleach" indiscriminately. By introducing variations among the cells so that they produced slightly different amounts of dye and bleach, Turing generated spots, dapples and even zebralike stripes. These slight initial differences were magnified by blotting and bleaching to create all-or-nothing patterns. We wondered if this notion would work for cortical maps.

Four years ago computational neuroscientist Misha Tsodyks and his colleagues at the Weizmann Institute of Science in Rehovot, Israel, demonstrated that, indeed, a similar process could generate cortexlike maps in software simulations. Paul Merolla, another doctoral student in my lab, took on the challenge of getting this self-organizing process to work in silicon. We knew that chemical dopants (impurities) introduced during the microfabrication process fell randomly, which introduced variations among otherwise identical transistors, so we felt this process could capture the randomness of gene expression in nature. That is putatively the source of variation of spot patterns from leopard to leopard and of orientation map patterns from person to person. Although the cells that create these patterns in nature express identical genes, they produce different amounts of the corresponding dye or ion channel proteins.

With this analogy in mind, Merolla designed a single silicon neuron and tiled it to create a mosaic with neuronlike excitatory and inhibitory connections among neighbors, which played the role of blotting and bleaching. When we fired up the chips in 2003, patterns of activity—akin to a leopard's spots—emerged. Different groups of cells became active when we presented edges with various orientations. By marking the locations of these different groups in different colors, we obtained orientation preference maps similar to those imaged in the V1 areas of ferret kits.

BUILDING BRAINS IN SILICON

Having morphed the retina's five layers into silicon, our goal turned to doing the same to all six of the visual cortex's layers. We have taken a first step by morphing layer IV, the cortex's input layer, to obtain an orientation preference map in an immature form. At three millimeters, however, the cortex is five times thicker than the retina, and morphing all six cortical layers requires integrated circuits with many more transistors per unit area.

Chip fabricators today can cram a million transistors and 10 meters of wire onto a square millimeter of silicon. By the end of this decade, chip density will be just a factor of 10 shy of cortex tissue density; the cortex has 100 million synapses and three kilometers of axon per cubic millimeter.

Researchers will come close to matching the cortex in terms of sheer numbers of devices, but how will they handle a billion transistors on a square centimeter of silicon? Thousands of engineers would be required to design these high-density nanotechnology chips using standard methods. To date, a hundredfold rise in design engineers accompanied the 10,000-fold increase in the transistor count in Intel's processors. In comparison, a mere doubling of the number of genes in flies to that of humans enabled evolutionary forces to construct brains with 10 million times more neurons. More sophisticated developmental processes made possible the increased complexity by elaborating on a relatively simple recipe. In the same way, morphing neural development processes instead of simply morphing neural circuitry holds great promise for handling complexity in the nanoelectronic systems of the future.

MORE TO EXPLORE

Mead, Carver. 1989. *Analog VLSI and neural systems.* Addison-Wesley.

Merolla, Paul, and Kwabena Boahen. 2004. A recurrent model of orientation maps with simple and complex cells. In *Advances in neural information processing systems,* vol. 16, ed. Sebastian Thrun, Larry Saul, and Bernhard Schölkopf. MIT Press.

Taba, Brian, and Kwabena Boahen. 2003. Topographic map formation by silicon growth cones. In *Advances in neural information processing systems,* vol. 15, ed. Suzanna Becker, Sebastian Thrun, and Klaus Obermayer. MIT press.

Zaghloul, Kareem A., and Kwabena Boahen. 2004. Optic nerve signals in a neuromorphic chip. *IEEE Transactions on Biomedical Engineering* 51 (4): 657–675.

KWABENA BOAHEN is a neuromorphic engineer and associate professor of bioengineering at the University of Pennsylvania. He left his native Ghana to pursue undergraduate studies in electrical and computer engineering at Johns Hopkins University in 1985 and became interested in neural networks soon thereafter. Boahen sees a certain elegance in neural systems that is missing in today's computers. He seeks to capture this sophistication in his silicon designs.

Originally published in *Scientific American,* vol. 292, No. 5, May 2005.

21

The Quest for a Smart Pill

New drugs to improve memory and cognitive performance in impaired
individuals are under intensive study. Their possible use in healthy people
already triggers debate

STEPHEN S. HALL

On a wintry afternoon in April, Tim Tully and I stood in a laboratory at Helicon Therapeutics, watching the future of human memory and cognition—or at least a plausible version of that future—take shape. Outside, a freak spring snowstorm lashed at the Long Island landscape. I mention the weather because it reminded both Tully and me of winters from our childhoods in the Midwest many years ago. The enduring power of those memories—and the biological proccsses that record and preserve them in the brain—lie at the heart of an incipient revolution in neuropharmacology that is unfolding in small, relatively unknown labs like this one in Farmingdale, N.Y.

Tully, a neuroscientist at Cold Spring Harbor Laboratory and founder of Helicon, has been one of the leading protagonists in the race to develop a new class of drugs that might improve memory in the memory-impaired—drugs that grow out of an increasingly sophisticated molecular and mechanistic understanding of how we can remember everything from snowstorms more than 30 years ago to where we put our car keys 30 minutes ago.

It is, alas, the nature of contemporary science (and commerce and bioethics, for that matter) that we often have to conjure up the future of human cognition, and its pharmacological manipulation, while staring at the behavior of a drugged mouse meandering in a jury-rigged box. So there we stood, gazing at a video playing on Tully's laptop computer, watching a small brown rodent enter an enclosed environment and begin its scurrying explorations in an experimental scenario known as Object Recognition Training. One day earlier, Tully

OVERVIEW/A BRAVE NEW PHARMACOLOGY
- An incipient revolution in neuropharmacology would offer drugs that could improve memory in those whose memories have faltered because of disease or aging and increase cognitive acuteness in fatigued individuals.
- Off-label use of some of these cognitive enhancers could allow normal individuals to sleep less, work harder and play more.
- Although most of these drugs are years away from government approval and clinical use, their possible social impact already has bioethicists contemplating the potential dangers.

explained, this same mouse had been placed in this same box, which contained two odd, knoblike objects, each with its distinct olfactory, tactile and other sensory tags. A mouse that is allowed to explore this environment for 15 minutes, Tully continued, will remember it so well that the animal will immediately notice any changes the next day; a mouse allowed to explore for only three and a half minutes, however, typically does not have enough time to commit the scene to long-term memory.

The mouse we were watching had had only three and a half minutes of training. But it did have a pharmaceutical assist, and that is what Tully wanted to show me. Narrating the action like a play-by-play announcer at a sports event, he described the scene as the little creature immediately paid an inordinate amount of murine attention to a new object in the room. "See, there he goes," Tully said in his earnest Midwestern locution. "He's walking around it. . . . Now he's climbing on top of it. He's not even paying attention to the other object." Indeed, the mouse sniffed at and circled and eventually clambered all over the novel object while ignoring the second object—the one encountered the day before.

To display this degree of curiosity, the mouse needed to *remember* what had been in the box the day before. That requires the formation of a long-term memory. And although years of behavioral experiments have established that mice ordinarily do not recall any changes in their environment after so brief a previous exposure, this one did, because of a drug—a memory drug known as a CREB enhancer—that Helicon hopes to begin testing in humans, perhaps as soon as the end of the year. "We've shown that several compounds will enhance the ability of a normal mouse to remember this task," Tully said. "And yet to make it a fact rather than a belief, we have to show it works in humans."

These days smart mice and erudite rats are the stalking-horses for a new pharmacology: drugs that might enhance human cognition, improving memory in those whose memories have faltered because of neurodegenerative disease or aging, perhaps even reengineering memory-forming circuitry in stroke victims or people with mental retardation. The potential market for such medicines is staggeringly large. As Tully and every other biotech and big-pharma executive know by heart, there are four million Americans with Alzheimer's disease, another 12 million with a condition called mild cognitive impairment (which often presages Alzheimer's), and approximately 76 million Americans older than 50, many of whom may soon satisfy a recent definition by the U.S. Food and Drug Administration for age-associated memory impairment (or AAMI), a form of mild forgetfulness. And judging by the sales of the herbal medicine ginkgo biloba, consumers are not waiting for an FDA-approved memory drug. Sales of ginkgo exceed $1 billion a year in the U.S., even though the scientific evidence that it improves memory is marginal at best; sales in Germany outstrip all acetylcholinesterase-inhibiting drugs used to slow memory loss in Alzheimer's patients, including donepezil (Aricept, marketed by Pfizer), rivastigmine (Exelon, marketed by Novartis) and galantamine (Reminyl, marketed by Janssen).

Despite an incessant media drumbeat about the coming revolution in what one magazine has dubbed "Viagra for the brain," smart pills are not around the corner. Cortex Pharmaceuticals in Irvine, Calif., has developed a class of memory-enhancing drugs called ampakines, which the company believes will increase the power of the neurotransmitter glutamate; the drugs have passed Phase I safety testing and are currently in Phase II tests (small-scale trials for efficacy) against Alzheimer's, mild cognitive impairment and schizophrenia. But those preliminary tests come at the end of a research odyssey that began in the mid-1980s, with no definitive end in sight.

Nevertheless, the action is beginning to heat up. Memory Pharmaceuticals in Montvale, N.J., which is commercializing the Nobel Prize–winning research of Columbia University professor Eric R. Kandel (see "The Biological Basis of Learning and Individuality" by Kandel, Eric R., and Robert D. Hawkins, *Scientific American*, September 1992), began initial safety testing of its first memory-enhancing drug in humans at the beginning of 2003, and Tully estimates that Helicon's lead drug candidate should enter trials no later than early 2004. Axonyx in New York City has been looking at phenserine (a potent acetylcholinesterase inhibitor) to treat Alzheimer's; the company began advanced testing in June. Princeton University neuroscientist Joe Z. Tsien, who caused an enormous stir in 1999 with the creation of a genetically enhanced smart mouse called Doogie, has advised a San Francisco–based biotech company, Eureka! Pharmaceuticals, which is collaborating with scientists in Shanghai to look for drugs that would merge modern genetics with ancient Chinese herbal medicine. Still, Tsien has his doubts about how soon the much-ballyhooed revolution will begin. "I'd be surprised to see any of these get to the clinic and become a drug anytime soon," he predicted, "especially a drug without side effects."

Although most of these new-generation drugs are years away from government approval and clinical use, their social impact has already been profound. Bioethicists have been working overtime contemplating the social dangers of memory enhancement, especially their potential use as "lifestyle" drugs. Moral philosopher Leon R. Kass, head of the President's Council on Bioethics, recently wrote that "in those areas of human life in which excellence has until now been achieved only by discipline and effort, the attainment of those achievements by means of drugs, genetic engineering, or implanted devices looks to be 'cheating' or 'cheap.'"

In another sense, however, the use of potent drugs as cognitive enhancers has been a feature of human life ever since people began drinking coffee. About 50 years ago the practice gained a more pharmaceutical aura when normal, healthy adults discovered that amphetamines could improve alertness. If, as some predict, the new cognitive enhancers are destined to replicate the pattern of Viagra and become lifestyle drugs, how might that happen, and how widespread might their use become? One possible answer may lie in an earlier generation of cognition-enhancing drugs that have already been approved— methylphenidate (Ritalin) for attentional focus, donepezil for Alzheimer's and modafinil for narcolepsy. These drugs are already taken by normal adults who seek to enhance mental acuity and performance. Users clearly believe that the drugs improve cognitive performance in normal people, although almost no research attests to this—and some research hints that they may be no better than a drug found on most breakfast tables.

THE CAFFEINE CAVEAT

Cognitive enhancement has been a feature of military research for a number of years. At Walter Reed Army Institute of Research, Nancy Jo Wesensten works on pharmaceutical agents that might improve the alertness (and therefore battlefield performance) of soldiers suffering severe sleep deprivation. In June 1998, while attending a meeting of sleep researchers, Wesensten stopped by the booth of Cephalon, a biotechnology company based in West Chester, Pa., and began chatting with one of its marketing representatives.

At the time, Cephalon was close to gaining FDA approval of a drug with the generic name of modafinil. Marketed as Provigil, this medicine is used to treat narcolepsy, the profound daytime drowsiness that afflicts an estimated 125,000 Americans. Modafinil, it became clear, would be an obvious candidate for the U.S. Army to test as a treatment for sleep deprivation—so much so that Wesensten was whisked up to the company's hospitality suite to discuss the work further. Eventually Cephalon agreed to provide modafinil for the army's research.

That was more than five years ago. In December 1998 the FDA approved the sale of modafinil in the U.S. to treat narcolepsy, and Cephalon is now selling about $200 million worth of the drug each year. That's a lot of narcolepsy medication—more, many observers suspect, than the U.S. population of narcoleptics can support. "There's a huge amount of off-label use by psychiatrists to augment mood," said Helene Emsellem, who runs the Center for Sleep and Wake Disorders in Chevy Chase, Md. In fact, modafinil is used to treat depression, multiple sclerosis and several other clinical conditions associated with fatigue. More to the point, there have been reports that doctors "are getting barraged" (as the online magazine *Slate* recently put it) by healthy people requesting prescriptions for modafinil as a cognitive enhancer that allows them to sleep less, stay up longer, work harder and play more. One well-known academic sleep researcher told me off the record, "People are telling me that they focus better on it, including some of my colleagues." Cephalon has been conducting clinical trials to test Provigil as a treatment for additional disorders of excessive sleepiness— resulting, for example, from disrupted sleep (caused by sleep apnea) or the "circadian misalignment" suffered by night-shift workers such as factory employees and truck drivers.

Which brings us back to Wesensten's study at Walter Reed's sleep center. "We were specifically interested in whether modafinil has any advantages over caffeine, which we find very good for reversing the effects of sleep deprivation on cognitive performance. Plus it's widely available, nonprescription and has a low side-effect profile," she said. "So was there any benefit to modafinil over caffeine?" Wesensten and her colleagues organized a randomized, double-blind, placebo study in which 50 volunteers were kept awake for 54 continuous hours. After about 40 hours, the subjects received either a placebo, 600 milligrams of caffeine (a stiff dose equal to about six cups of coffee) or one of three doses of modafinil (100 milligrams, 200 milligrams or 400 milligrams). Then they were subjected to a battery of tests to assess cognitive function and side effects.

The bottom line? The highest dose of modafinil, 400 milligrams, cut through fatigue and restored cognitive performance to normal levels—but so

HOW SOME MEMORY DRUGS WOULD WORK

Certain memory drugs under study influence two processes that operate when neurons encode long-term memories: membrane depolarization and activation of the CREB protein.

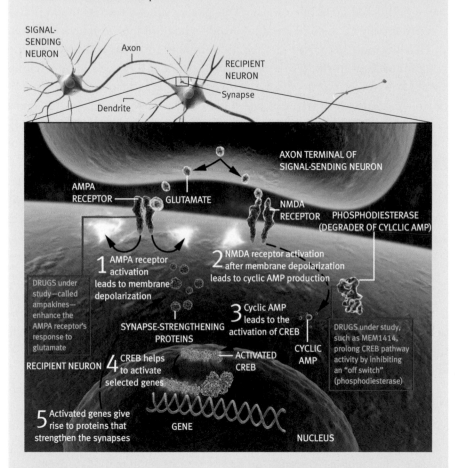

Depolarization can occur after release of the excitatory neurotransmitter glutamate at synapses (contact points between two nerve cells) stimulates AMPA receptors on recipient cells. Depolarization, however it happens, helps another surface protein, the NMDA receptor, to respond to glutamate. The receptor reacts by activating the CREB pathway inside cells—a series of molecular interactions that includes production of a molecule called cyclic

AMP, which leads to activation of CREB. (Broken arrows indicate that steps in the pathway have been omitted for simplicity.) This last event is key: activated CREB helps switch on genes whose protein products strengthen specific synapses.

Some drugs under investigation aim to speed memory storage by amplifying the AMPA receptor's response to glutamate and thus facilitating depolarization. Other compounds aim to increase a cell's supply of active CREB—such as by inhibiting an enzyme (phosphodiesterase) that normally degrades the cyclic AMP needed for CREB activation.

did caffeine. The reported side effects of modafinil were quite low—but so were those of caffeine. "What we concluded," Wesensten said, "was that there didn't appear to be any benefit to using modafinil over caffeine. It just wasn't there. Both drugs looked very similar."

The U.S. Air Force has also conducted extensive experimentation with drugs that increase alertness in fatigued military personnel, a particular concern for pilots in an operational setting. The air force allowed use of amphetamines as "go pills" by pilots as early as World War II, according to John A. Caldwell, a sleep disorders expert with the air force who has conducted such experiments over the past 10 years. "My primary objective is not to enhance cognitive performance," he said in an interview, "but to maintain the already excellent performance levels of our military."

Beginning in 1993, Caldwell carried out randomized, double-blind experiments showing that dextroamphetamine eliminates virtually all the decrements of performance in both male and female pilots who have not slept for 40 hours. Some of the studies took place in a helicopter flight simulator but have been replicated in real aircraft. More recently, he tested modafinil head-to-head against dextroamphetamine in sleep-deprived pilots, showing that the narcolepsy drug overcame fatigue and maintained cognitive performance, although some of the subjects developed nausea akin to motion sickness inside the simulator. "Ultimately, I think there will be a place for modafinil," Caldwell said. "It wouldn't surprise me if it would be approved for use within a year. But I don't think it will be a replacement for our current 'go pill.' We have 50 years of operational experience, and tons of laboratory research, with dextroamphetamine. We're not there yet with modafinil."

A HALO OF POWDER

Research on modafinil, nonetheless, highlights a paradox in the ethical debate about cognitive enhancement. The Defense Advanced Research Projects Agency (DARPA) has funded considerable basic and clinical research looking at ways for its personnel to increase cognitive performance. Its Continuous Assisted Performance (CAP) program has funded preclinical research with Cortex's ampakine drugs, for example. So whereas members of one government body, President George W. Bush's bioethics panel, have characterized the use of drugs by healthy people to enhance cognitive performance as a form of cheating, another branch of the government, the military, has aggressively explored the capacity of new pharmaceutical agents to increase cognitive alertness and performance in fatigued but essentially normal individuals—a short hop, skip and a jump to cognitive enhancement for civilians.

Modafinil is merely the latest cognitive enhancer to develop a following among healthy individuals. There is a mini-literature (not to say mythology) surrounding the use of Ritalin as a study aid by high school and college students. Ritalin, marketed by Novartis, is typically prescribed for children with attention-deficit hyperactivity disorder (ADHD) but has reportedly found favor with students and even business executives. Several students at a prestigious East Coast preparatory school told me that Ritalin use as a study aid was so common that students occasionally sported a halo of powder around their nostrils after snorting the drug. The practice has spread to college campuses. "It's here," confirmed Eric Heiligenstein, clinical director of psychiatry at the University of Wisconsin Health Services. "It's fairly well established, if you want to use it." Although the amount of Ritalin consumed by college students is almost impossible to quantify, Heiligenstein said that the number of hard-core users is "very small" yet more extensive than those who take modafinil because Ritalin is "available, relatively cheap and has a pretty good safety profile."

Among the sparse findings about the effects of these drugs on healthy individuals, at least one study suggests that a long-standing dementia treatment improves cognitive functioning in normal people. In July 2002 Jerome A. Yesavage of Stanford University, Peter J. Whitehouse of Case Western Reserve University and their colleagues published a study in *Neurology* assessing the impact of donepezil on the performance of pilots. Donepezil, marketed as Aricept, is one of many drugs approved by the FDA to slow the progressive memory loss experienced by patients with Alzheimer's disease. The researchers trained two groups of pilots in a Cessna 172 flight simulator; one group then received a placebo while the other group took five milligrams of donepezil, less than the

STATE OF THE ART FOR SMART

COGNITIVE ENHANCEMENT drugs, some of which are still under development, focus so far on treating dementia and other disorders. Some compounds on the market are also being used or tested to improve normal functioning, such as to increase wakefulness in shift workers or to help pilots perform under stress.

TYPE OF DRUG	COMPANY	PURPOSE	STATUS*
CREB suppressor	Helicon Therapeutics	Suppression of disturbing memories	Early stages of development
CREB enhancer	Helicon Therapeutics	Memory enhancement	Early stages of development
CREB enhancer (MEM 1414)	Memory Pharmaceuticals in partnership with Roche	Memory enhancement	Will enter Phase 1 trials in late 2003
Calcium flow regulator (MEM 1003)	Memory Pharmaceuticals	Memory enhancement	In Phase I trials
Ampakines	Cortex Pharmaceuticals	Memory enhancement	In Phase II trials
Phenserine	Axonyx	Treatment of mild to moderate Alzheimer's	Phase II trials completed
Modafinil (Provigil)	Cephalon	Treatment of narcolepsy	On the market
Methylphenidate (Ritalin)	Novartis	Attention enhancement	On the market
Donepezil (Aricept)	Eisai/Pfizer	Treatment of mild to moderate Alzheimer's	On the market
Rivastigmine (Exelon)	Novartis	Treatment of mild to moderate Alzheimer's	On the market
Galantamine (Reminyl)	Janssen	Treatment of mild to moderate Alzheimer's	On the market

*Phase I trials study the safety of a new drug in small, healthy human populations. Phase II trials examine safety and efficacy in individuals afflicted with the disorder in question. To gain approval, drugs must also pass through large, Phase III, trials of safety and efficacy.

routine dose for Alzheimer's, for 30 days. Then they tested both groups again in the simulator.

Yesavage and his colleagues threw several curves at the pilots—they were asked to perform some complicated air-traffic maneuvers and had to react to in-flight emergencies, including a drop in oil pressure as indicated by cockpit instrumentation. A month after their initial training, the pilots on donepezil performed significantly better than the control group, with especially enhanced performance on the landing approach and in handling emergencies. Yesavage, who hopes to conduct an expanded study sometime soon, noted in the *Neurology* article that "if cognitive enhancement becomes possible in intellectually intact individuals, significant legal, regulatory, and ethical questions will emerge."

If those questions are true of donepezil, modafinil and other existing drugs, they will be especially true for the new generation of smart drugs, precisely because they are based on a mechanistic approach to memory that could be particularly powerful—unlike the accidental discoveries we have often had up to now. And although every biotech executive decries the notion of a lifestyle drug, everyone is aware of the precedent. "Typically industry wanted to avoid enhancement drugs in the 1990s," said one neuroscientist. "But I think Viagra changed a lot of people's opinion."

IMPROVING MEMORY

As he guided me through some 32,000 square feet of drug-discovery real estate at Memory Pharmaceuticals in northern New Jersey, Axel Unterbeck punctuated every stop on the tour with the phrase "very sophisticated." Unterbeck, the company's tall, charming, elegantly dressed president and chief scientific officer, invoked the words again and again—in the electrophysiology lab where half a dozen biologists record the effect of potential memory-enhancing drugs on individual neurons and slices of animal brain, in the vivarium where the company tests those candidate drugs in elderly rodents, and in the pharmacokinetics room, where the disembodied whines and clicks of robotic machinery accompany the analysis of blood samples from animals and humans. "They're doing the job as we speak," Unterbeck said, proudly pointing out a $250,000 machine that speedily determines the concentration of drug metabolites in blood. "*Very* sophisticated."

Everything about Memory Pharmaceuticals bespeaks state-of-the-art science and high-end ambition—its intellectual godfathers and founders (Columbia Nobel laureate Kandel and Harvard Nobel laureate Walter Gilbert), its beautifully landscaped headquarters with birch trees and daffodils flanking the entryway, even its tony neighbors (the North American headquarters of Mercedes-Benz is just up the road). Founded in 1998, the company is betting a lot of money—$41.5 million from a recent round of financing, plus a co-development deal potentially worth $150 million with the Swiss drug giant Roche—that it can navigate the shoals of drug discovery more efficiently by identifying toxicological and pharmacokinetic (drug metabolism) problems in cognition-enhancing drugs early in the process. "That's the future," Unterbeck said, "and we are very well positioned for translating the science into smart drugs."

Early in 2003 Memory Pharmaceuticals began initial safety testing of its first smart drug, a compound called MEM 1003, in healthy volunteers in London. The compound regulates the flow of calcium ions into neurons and is designed to restore the equilibrium of calcium in brain cells that have been disrupted by Alzheimer's, mild cognitive impairment or a condition called vascular dementia. "So far this program looks exceptionally good in terms of pharmacokinetics and toxicology," Unterbeck said. "The compound looks exceptionally safe." But perhaps the most closely watched of Memory's potential smart drugs is a compound called MEM 1414, because this drug would tweak a molecular pathway identified by Kandel's and Tully's labs as crucial to converting short-term experience and learning into long-term memory. It involves a powerful protein known as CREB.

In the mid-1990s Tully and Jerry Yin of Cold Spring Harbor Laboratory genetically engineered a fruit fly that displayed the insect equivalent of a photographic memory—these flies learned and memorized a task after one training exercise, whereas normal flies took 10 practice sessions. They managed this stunning enhancement of memory by goosing the output of a single gene called CREB. Both Tully's and Kandel's labs have shown that when simple animals learn a task and commit it to memory, the synapses used to form the memory are remodeled and strengthened in a process that requires the activation of genes. As it turns out, memory formation unleashes a messenger molecule inside the cell known as cyclic adenosine monophosphate, or AMP. This molecule in turn triggers the formation of a protein that binds to the DNA of a nerve cell, thus activating an entire suite of genes that add the mortar and brick at synapses to consolidate memory formation. This instigating protein is called cyclic AMP response element-binding protein, or CREB. The more CREB swimming around in a neuron, the faster long-term memory is consolidated. That at least has been the case with sea mollusks, fruit flies and mice. Now the question is: Will it be true of humans, too?

Normally, another chemical—phosphodiesterase (PDE)—breaks down cyclic AMP in the cell. Pharmacologically inhibiting phosphodiesterase makes more CREB available for a longer period—thus, in theory, strengthening and speeding the process of memory formation. Phosphodiesterase inhibitors have a spotty reputation in pharmaceutical circles, however; one version was approved in Japan to treat depression, but the molecules have a history of causing nausea. Nevertheless, PDE inhibitors have performed exceedingly well as memory enhancers in preclinical testing, according to researchers in the field, because they allow more CREB to hang around in the cell during learning, which promotes memory consolidation. Hence, both Memory Pharmaceuticals and Helicon Therapeutics are developing drugs based on a class of molecules known as PDE-4. Helicon is also working on a drug that suppresses memories, something that might be used to block or erase disturbing memories of a traumatic event. "We have preclinical evidence that suggests that they might selectively block traumatic memories that have formed before," Tully said.

Memory Pharmaceuticals is especially high on its MEM 1414 molecule—a fascination ratified in July 2002 when Roche agreed to be a partner in its development. "What is really interesting, you see the same kind of age-associated memory impairment in nonhuman primates and rodents as you see in humans," Unterbeck explained. About 50 percent of aged animals, he continued, are unable to form new memories, yet MEM 1414 restored age-related deficits in the animals' recall to close to normal. The company launched Phase I tests (for safety) of the compound this year.

Even an ideal progression through clinical testing and federal drug approval, however, adverts to a slow and perilous timeline. "MEM 1003 could—and it's a big could—be on the market in 2008," said Tony Scullion, Memory's CEO, "and 1414 wouldn't be too far behind." But as Unterbeck knows from his previous tenure at Bayer, the promise of a new drug often doesn't unravel until late in the game, when the large number of patients typically enrolled in Phase III trials can reveal not only less-than-optimal efficacy but more-than-expected side effects. "Drug companies put $500 million down," he said, "and you get failure in Phase III." Larry Squire, an elder statesman of memory research at the University of California at San Diego, added, "In fact, you could say the whole history of the field has been to deal with side effects."

Moreover, there is hardly unanimity that CREB is the best or only route to a blockbuster memory drug. "There's not very strong biology in the CREB pathway, especially in mammalian systems," one neuroscientist who requested anonymity pointed out. "The targets are not well validated, and CREB is expressed everywhere, very early on." Another prominent neuroscientist told me that even a scientific adviser to Memory Pharmaceuticals has privately expressed the view that the new drugs may prove no more effective than caffeine. Nor is CREB the only portal to memory manipulation. Tsien, creator of the smart mouse at Princeton, is pursuing a different memory pathway involving a receptor of the neurotransmitter NMDA that is limited to the forebrain; and Cortex's ampakine technology is focused on a different neurotransmitter system. "Frankly speaking, we still know so little," Tsien said. "We know no principles, no operating code for memory. We know a lot of genes, but we don't have a coherent picture, and I think that is the problem with the whole area of therapeutic research and development."

Researchers are resigned to the continuing bioethical debate on the drugs, no matter how premature the science or how fuzzy the future. "We've got our hands full just showing that these drugs will work," admitted Tully, who has a long history of being keenly attentive to the social implications of scientific research. "Having said that, do I think there will be off-label use if it works clinically? Yes, I do. In principle, these compounds could improve the motor skills required to play the piano or second-language acquisition. The off-label use of drugs happened with Viagra, and it didn't stop Viagra, it didn't stop Ritalin, and it didn't stop amphetamines. But the fact is, off-label use of prescription drugs is dangerous because of unanticipated side effects. You may create unknown psychological problems. But it's not worth even talking about at all except as science fiction. We simply have to wait until we put these drugs into people and see what happens."

Given that we are most likely five or 10 years away from "seeing what happens," we're probably destined to read a lot more about smart drugs before we actually have any pills in hand. But there may be a cautionary warning in a little episode that happened when I visited Tsien at Princeton. He was walking me through the animal facility, which houses his genetically engineered "smart" mice, when one of the lab technicians walked by holding a mouse trap with two unhappy occupants. Tsien looked down at the two cognitively enhanced rodents in the trap, shook his head and said simply, "Not so smart."

MORE TO EXPLORE

Kandel, Eric. Nobel address. Available online at http://nobelprize.org/nobel_prizes/medicine/laureates/2000/kandel-speech.html

Remembering and forgetting. Sessions 3 and 4 of the President's Council on Bioethics, October 17, 2002. Available online at www.bioethics.gov/transcripts/oct02

Squire, Larry R., and Eric Kandel. 1999. *Memory: From mind to molecules*. Scientific American Library, no. 69. W. H. Freeman.

Tully, Tim, Rusiko Bourtchouladze, Rod Scott, and John Tallman. 2003. Targeting the CREB pathway for memory enhancers. *Nature Reviews Drug Discovery* 2 (4): 267–277.

STEPHEN S. HALL is a writer based in New York City. He has written four books chronicling the contemporary history of science, including most recently an account of stem cell and cloning research, *Merchants of Immortality* (Houghton Mifflin, 2003). Originally published in *Scientific American*, Vol. 289, No. 3, September 2003.

INDEX

A

AAMI (age-associated memory impairment), 235, 244

acetylcholine, 4, 43, 235

Ackermann, Hermann, 93

acoustic-temporal changes, 103

action potentials, 197–198, 201–203, 207, 214–215

activation negativity, 214, 215

acute phase of depression, 180, 181, 184–187

AD. *See* Alzheimer's disease

addiction
    biological process, 143–146
    and CREB protein, 145, 147, 149–154
    Freudian libido and, 42
    medical illnesses from, 154
    to new discoveries, 75
    overview, 142–143
    similar response to variety of stimuli, 146–147, 152
    social and emotional factors, 154–155
    stages of, 147–151
    tolerance development, 151–154
    and VTA-accumbens pathway, 146
    See also nucleus accumbens neurons; reward pathways

adenosine monophosphate (AMP), 73, 144–145, 238–239, 243–244

ADHD (attention-deficit hyperactivity disorder), 95, 240

advance directives, 57

age-associated memory impairment (AAMI), 235, 244

agoraphobia, 139

alcohol abuse, 152–153, 164. *See also* addiction

alpha 7 nicotine receptor stimulators, 163

ALS (Lou Gehrig's disease), 119, 127, 128, 213

Alzheimer's disease (AD)
    definitive signs of, 134–135
    diagnosing with fMRI, 140
    and donepezil, 240, 242
    and neurogenesis, 127
    and neuropharmacology, 234, 235–236
    overview, 56–57

amacrine cells, 226–227

ambivalence, 158

American Psychiatric Association, 132

American Sign Language (ASL), 103, 106–107, 111

amino acids, 4, 115, 165–167. *See also* glutamate

amnesia, 40, 49, 51, 54

amoeboid structure on axons, 228

AMP (adenosine monophosphate), 73, 144–145, 238–239, 243–244

AMPA receptors, 238–239

ampakine drugs, 163, 235, 238–239, 240, 241

AMPA-type glutamate receptor stimulators, 163, 166

amphetamines, 152–153, 158–159, 236. *See also* addiction

amygdala, 3, 146, 148, 151–154

amyloid precursor protein, 127

amyloid protein, 134–135

amyotrophic lateral sclerosis (ALS), 119, 127, 128, 213

anatomical abnormalities with schizophrenia, 138, 140

anatomical position of regions in the cortex, 4

Andreasen, Nancy C., 97

Andrich, Juergen, 114–120

angel dust, 162

anger-rage instinctual system, 42

animal drives of humans, Freud on, 40, 42

anosognosic patients, 37, 38

anterior insula, 50, 53

antianxiety drugs, 134

antidepressant drugs, 119, 134, 179, 182–187

antioxidants, 119

antipsychotic drugs, 119, 140, 156–157, 161

anxiety disorders, 139

aphasias, 102–103, 104

Aplysia (sea slug), 73–74

apoptosis, 3, 118

Archimedes, 18

Aricept (donepezil), 236, 240, 241, 242

OTHER DANA PRESS BOOKS
AND PERIODICALS

www.dana.org/books/press

BOOKS FOR GENERAL READERS
BRAIN and MIND

CEREBRUM 2007: Emerging Ideas in Brain Science
Cynthia A. Read, Editor · Foreword by Bruce McEwen, Ph.D.
Prominent scientists and other thinkers explain, applaud, and protest new
ideas arising from discoveries about the brain in this first yearly anthology
from *Cerebrum's* Web journal for inquisitive general readers. Visit *Cerebrum*
online at www.dana.org/cerebrum. 10 black-and-white illustrations.
PAPER · 225 PP · 1-932594-24-8 · $14.95

MIND WARS: Brain Research and National Defense
Jonathan Moreno, Ph.D.
A leading ethicist examines national security agencies' work on defense
applications of brain science, and the ethical issues to consider.
CLOTH · 210 PP · 1-932594-16-7 · $23.95

THE DANA GUIDE TO BRAIN HEALTH: A Practical Family
Reference from Medical Experts (with CD-ROM)
Floyd E. Bloom, M.D., M. Flint Beal, M.D., and David J. Kupfer, M.D., Editors
Foreword by William Safire
The only complete, authoritative, family-friendly guide to the brain's devel-
opment, health, and disorders. *The Dana Guide to Brain Health* offers ready
reference to our latest understanding of brain diseases as well as information
to help you participate in your family's care. 16 full-color pages and more than
200 black-and-white drawings.
PAPER (WITH CD-ROM) · 733 PP · 1-932594-10-8 · $25.00

THE CREATING BRAIN: The Neuroscience of Genius
Nancy C. Andreasen, M.D., Ph.D.
A noted psychiatrist and bestselling author explores how the brain achieves cre-
ative breakthroughs, including questions such as how creative people are differ-

ent and the difference between genius and intelligence. She also describes how
to develop our creative capacity. 33 illustrations/photos.

CLOTH · 197 PP · 1-932594-07-8 · $23.95

THE ETHICAL BRAIN
Michael S. Gazzaniga, Ph.D.

Explores how the lessons of neuroscience help resolve today's ethical dilem-
mas, ranging from when life begins to free will and criminal responsibility.
The author, a pioneer in cognitive neuroscience, is a member of the President's
Council on Bioethics.

CLOTH · 201 PP · 1-932594-01-9 · $25.00

A GOOD START IN LIFE: Understanding Your Child's Brain and Behavior from Birth to Age 6
Norbert Herschkowitz, M.D., and Elinore Chapman Herschkowitz

The authors show how brain development shapes a child's personality and
behavior, discussing appropriate rule-setting, the child's moral sense, tem-
perament, language, playing, aggression, impulse control, and empathy.
13 illustrations.

CLOTH · 283 PP · 0-309-07639-0 · $22.95 | PAPER (UPDATED WITH NEW MATERIAL) · 312 PP · 0-9723830-5-0 · $13.95

BACK FROM THE BRINK: How Crises Spur Doctors to New Discoveries about the Brain
Edward J. Sylvester

In two academic medical centers, Columbia's New York Presbyterian and Johns
Hopkins Medical Institutions, a new breed of doctor, the neurointensivist,
saves patients with life-threatening brain injuries. 16 illustrations/photos.

CLOTH · 296 PP · 0-9723830-4-2 · $25.00

THE BARD ON THE BRAIN: Understanding the Mind Through the Art of Shakespeare and the Science of Brain Imaging
Paul Matthews, M.D., and Jeffrey McQuain, Ph.D. · Foreword by Diane Ackerman

Explores the beauty and mystery of the human mind and the workings of
the brain, following the path the Bard pointed out in 35 of the most famous
speeches from his plays. 100 illustrations.

CLOTH · 248 PP · 0-9723830-2-6 · $35.00

STRIKING BACK AT STROKE: A Doctor-Patient Journal
Cleo Hutton and Louis R. Caplan, M.D.

A personal account with medical guidance from a leading neurologist for anyone enduring the changes that a stroke can bring to a life, a family, and a sense
of self. 15 illustrations.

CLOTH · 240 PP · 0-9723830-1-8 · $27.00

UNDERSTANDING DEPRESSION: What We Know
and What You Can Do About It
J. Raymond DePaulo Jr., M.D., and Leslie Alan Horvitz · Foreword by Kay Redfield Jamison, Ph.D.
What depression is, who gets it and why, what happens in the brain, troubles
that come with the illness, and the treatments that work.
CLOTH · 304 PP · 0-471-39552-8 · $24.95 | PAPER · 296 PP · 0-471-43030-7 · $14.95

KEEP YOUR BRAIN YOUNG: The Complete Guide to Physical
and Emotional Health and Longevity
Guy McKhann, M.D., and Marilyn Albert, Ph.D.
Every aspect of aging and the brain: changes in memory, nutrition, mood,
sleep, and sex, as well as the later problems in alcohol use, vision, hearing,
movement, and balance.
CLOTH · 304 PP · 0-471-40792-5 · $24.95 | PAPER · 304 PP · 0-471-43028-5 · $15.95

THE END OF STRESS AS WE KNOW IT
Bruce McEwen, Ph.D., with Elizabeth Norton Lasley · Foreword by Robert Sapolsky
How brain and body work under stress and how it is possible to avoid its
debilitating effects.
CLOTH · 239 PP · 0-309-07640-4 · $27.95 | PAPER · 262 PP · 0-309-09121-7 · $19.95

IN SEARCH OF THE LOST CORD: Solving the Mystery of Spinal Cord Regeneration
Luba Vikhanski
The story of the scientists and science involved in the international scientific
race to find ways to repair the damaged spinal cord and restore movement.
21 photos; 12 illustrations.
CLOTH · 269 PP · 0-309-07437-1 · $27.95

THE SECRET LIFE OF THE BRAIN
Richard Restak, M.D. · Foreword by David Grubin
Companion book to the PBS series of the same name, exploring recent discov-
eries about the brain from infancy through old age.
CLOTH · 201 PP · 0-309-07435-5 · $35.00

THE LONGEVITY STRATEGY: How to Live to 100 Using the Brain-Body Connection
David Mahoney and Richard Restak, M.D. · Foreword by William Safire
Advice on the brain and aging well.
CLOTH · 250 PP · 0-471-24867-3 · $22.95 | PAPER · 272 PP · 0-471-32794-8 · $14.95

STATES OF MIND: New Discoveries about How Our Brains Make Us Who We Are
Roberta Conlan, Editor

Adapted from the Dana/Smithsonian Associates lecture series by eight of the country's top brain scientists, including the 2000 Nobel laureate in medicine, Eric Kandel.

CLOTH · 214 PP · 0-471-29963-4 · $24.95 | PAPER · 224 PP · 0-471-39973-6 · $18.95

THE DANA FOUNDATION SERIES ON NEUROETHICS

DEFINING RIGHT AND WRONG IN BRAIN SCIENCE:
Essential Readings in Neuroethics
Walter Glannon, Ph.D., Editor

The fifth volume in The Dana Foundation Series on Neuroethics, this collection marks the five-year anniversary of the first meeting in the field of neuroethics, providing readers with a complete guide to the historical, present, and future ethical issues facing neuroscience and society. 10 black-and-white illustrations.

PAPER · 350 PP · 1-932594-25-6 · $15.95

HARD SCIENCE, HARD CHOICES: Facts, Ethics, and Policies
Guiding Brain Science Today
Sandra Ackerman, Editor

Top scholars and scientists discuss new and complex medical and social ethics brought about by advances in neuroscience. Based on an invitational meeting co-sponsored by the Library of Congress, the National Institutes of Health, the Columbia University Center for Bioethics, and the Dana Foundation.

PAPER · 152 PP · 1-932594-02-7 · $12.95

NEUROSCIENCE AND THE LAW: Brain, Mind, and the Scales of Justice
Brent Garland, Editor. With commissioned papers by Michael S. Gazzaniga, Ph.D., and Megan S. Steven; Laurence R. Tancredi, M.D., J.D.; Henry T. Greely, J.D.; and Stephen J. Morse, J.D., Ph.D.

How discoveries in neuroscience influence criminal and civil justice, based on an invitational meeting of 26 top neuroscientists, legal scholars, attorneys, and state and federal judges convened by the Dana Foundation and the American Association for the Advancement of Science.

PAPER · 226 PP · 1-932594-04-3 · $8.95

BEYOND THERAPY: Biotechnology and the Pursuit of Happiness.
A Report of the President's Council on Bioethics
Special Foreword by Leon R. Kass, M.D., Chairman · Introduction by William Safire

Can biotechnology satisfy human desires for better children, superior performance, ageless bodies, and happy souls? This report says these possibilities present us with profound ethical challenges and choices. Includes dissenting commentary by scientist members of the Council.

PAPER · 376 PP · 1-932594-05-1 · $10.95

NEUROETHICS: Mapping the Field. Conference Proceedings.
Steven J. Marcus, Editor
Proceedings of the landmark 2002 conference organized by Stanford University and the University of California, San Francisco, and sponsored by The Dana Foundation, at which more than 150 neuroscientists, bioethicists, psychiatrists and psychologists, philosophers, and professors of law and public policy debated the ethical implications of neuroscience research findings. 50 illustrations.

PAPER · 367 PP · 0-9723830-0-X · $10.95

IMMUNOLOGY

RESISTANCE: The Human Struggle Against Infection
Norbert Gualde, M.D., translated by Steven Rendall
Traces the histories of epidemics and the emergence or re-emergence of diseases, illustrating how new global strategies and research of the body's own weapons of immunity can work together to fight tomorrow's inevitable infectious outbreaks.

CLOTH · 219 PP · 1-932594-00-0 $25.00

FATAL SEQUENCE: The Killer Within
Kevin J. Tracey, M.D.
An easily understood account of the spiral of sepsis, a sometimes fatal crisis that most often affects patients fighting off nonfatal illnesses or injury. Tracey puts the scientific and medical story of sepsis in the context of his battle to save a burned baby, a sensitive telling of cutting-edge science.

CLOTH · 231 PP · 1-932594-06-X · $23.95 | PAPER · 231 PP · 1-932594-09-4 · $12.95

ARTS EDUCATION

A WELL-TEMPERED MIND: Using Music to Help Children Listen and Learn
Peter Perret and Janet Fox · Foreword by Maya Angelou
Five musicians enter elementary school classrooms, helping children learn about music and contributing to both higher enthusiasm and improved academic performance. This charming story gives us a taste of things to come in one of the newest areas of brain research: the effect of music on the brain. 12 illustrations.

CLOTH · 225 PP · 1-932594-03-5 · $22.95 | PAPER · 225 PP · 1-932594-08-6 · $12.00

FREE EDUCATIONAL BOOKS
(Information about ordering and downloadable PDFs are available at www.dana.org.)

PARTNERING ARTS EDUCATION: A Working Model from ArtsConnection
This publication describes how classroom teachers and artists learned to form partnerships as they built successful residencies in schools. *Partnering Arts Education* provides insight and concrete steps in the ArtsConnection model. 55 PP

ACTS OF ACHIEVEMENT: The Role of Performing Arts Centers in Education
Profiles of more than 60 programs, plus eight extended case studies, from urban and rural communities across the United States, illustrating different approaches to performing arts education programs in school settings. Black-and-white photos throughout. 164 PP

PLANNING AN ARTS-CENTERED SCHOOL: A Handbook
A practical guide for those interested in creating, maintaining, or upgrading arts-centered schools. Includes curriculum and development, governance, funding, assessment, and community participation. Black-and-white photos throughout. 164 PP

THE DANA SOURCEBOOK OF BRAIN SCIENCE: Resources for
Teachers and Students, Fourth Edition
A basic introduction to brain science, its history, current understanding of the brain, new developments, and future directions. 16 color photos; 29 black-and-white photos; 26 black-and- white illustrations. 160 PP

THE DANA SOURCEBOOK OF IMMUNOLOGY: Resources for Secondary
and Post-Secondary Teachers and Students
An introduction to how the immune system protects us, what happens when it breaks down, the diseases that threaten it, and the unique relationship between the immune system and the brain. 5 color photos; 36 black-and-white photos; 11 black-and-white illustrations. 116 PP · ISSN: 1558-6758

PERIODICALS

Dana Press also offers several periodicals dealing with arts education, immunology, and brain science. These periodicals are available free to subscribers by mail. Please visit www.dana.org.

CREDITS

Pg. 10: Courtesy of Jancy Chang

Pg. 11: Courtesy of Jancy Chang

Pg. 14: SIGANIM, Gehirn & Geist

Pg. 16: Erich Lessing/Art Resource, NY

Pg. 23: Bryan Christie Design

Pg. 25: Bryan Christie Design

Pg. 27: Bryan Christie Design

Pg. 38: Reprinted with permission from the estate of A.W. Freud et al., by arrangement with Paterson Marsh Ltd., (a); Oliver Turnbull (b, all coloring)

Pg. 41: (c) Bettman/CORBIS (photograph); Reprinted with permission from the estate of A.W. Freud et al., by arrangement with Paterson Marsh Ltd. (drawing)

Pg. 52: Reprinted by permission from Macmillan Publishers Ltd: *Nature Neuroscience*, Vol. 5, No. 9, David J. Turk et al., "Mike or Me? Self-Recognition in a Split-Brain Patient," 2002

Pg. 53: 3FX, Inc.

Pg. 61: Hanna Damasio

Pg. 63: Dimitry Schidlovsky; Source: Tootell et al., in *Journal of Neuroscience*, May 1988

Pg. 80: Old Age, Adolescence, Infancy (The Three Ages) (1940) Oil on canvas, 19 5/8 x 25 5/8 inches. (c) Salvador Dalí. Fundación Gala-Salvador Dalí, (Artists Rights Society), 2007. Collection of the Salvador Dalí Museum, Inc., St. Petersburg, FL, 2007.

Pg. 81: Johnny Johnson

Pg. 84: Terese Winslow, with assistance from Heidi Baseler, Bill Press, and Brian Wandell, Stanford University

Pg. 86: Nikos K. Logothetis

Pg. 91: Alice Y. Chen; Source: David C. Van Essen, Washington University School of Medicine

Pg. 94: Alice Y. Chen

Pg. 98-99: Matt Collins (illustrations); Reprinted from Jia-Hong Gao et al. in *Science*, Vol. 272; April 26, 1996 (brain scans)

Pg. 104: Illustration by Peter Stemler

Pg. 106-107: Illustrations by Peter Stemler

Pg. 118: SIGANIM, Gehirn & Geist

Pg. 123: Alice Y. Chen

Pg. 138: Thompson /Toga-UCLA and Rapoport/Geidd-NIMH

Pg. 144-145: Terese Winslow

Pg. 148: Hans C. Breiter, Massachusetts General Hospital (scans and graph)

Pg. 149: Reprinted from *Neuroscience*, Vol. 116, S.D. Norrholm et al., Cocaine-induced proliferation of dendritic spines in nucleus accumbens is dependent on the activity of cyclin-dependent kinase-5. Pages 19-22, 2003, with permission from Elsevier (micrographs)

Pg. 152-153: Terese Winslow

Pg. 159: iStockPhoto.com

Pg. 160-161: Alfred T. Kamajian

Pg. 175: Courtesy of Helen Mayberg

Pg. 180: Source: Steven Hollon et al. in *Psychological Science in the Public Interest*, Vol. 3, No. 2; November 2002

Pg. 181: Source: E. Frank et al. in *Archives of General Psychiatry*, Vol. 47; 1990

Pg. 186: Source: M. B. Keller et al. in *New England Journal of Medicine*, Vol. 342; 2000

Pg. 189: Laurie Grace

Pg. 194: Edmond Alexander

Pg. 198: Duke Photography

Pg. 204-205: Bryan Christie Design

Pg. 206: Bryan Christie Design

Pg. 214: Niels Birbaumer et al. in *IEEE Transactions on Neural Systems and Rehabilitation Engineering*, Vol. 11, No. 2; June 2003

Pg. 216: Courtesy of Jose del R. Millan, Dalle Molle Institute for Perceptual Artificial Intelligence

Pg. 221: Courtesy of Kareem Zaghloul

Pg. 226-227: Bryan Christie Design; Kareem Zaghloul (silicon retina)

Pg. 238: Mike Medicine Horse, *Hybrid Medical Animation*

"The New Science of Mind" reprinted with permission from Eric R. Kandel. All other articles reprinted with permission from *Scientific American*.